LOGICISM AND
THE PHILOSOPHY OF LANGUAGE

LOGICISM AND THE PHILOSOPHY OF LANGUAGE

Selections from Frege and Russell

edited by
Arthur Sullivan

broadview press

©2003 Arthur Sullivan

All rights reserved. The use of any part of this publication reproduced, transmitted in any form or by any means, electronic, mechanical, photocopying, recording, or otherwise, or stored in a retrieval system, without prior written consent of the publisher — or in the case of photocopying, a licence from ACCESS COPYRIGHT (Canadian Copyright Licensing Agency), One Yonge Street, Suite 1900, Toronto, Ontario M5E 1E5 — is an infringement of the copyright law.

National Library of Canada Cataloguing in Publication

Sullivan, Arthur, 1970-
  Logicism and the philosophy of language: selections from Frege and Russell / Arthur Sullivan.

Includes bibliographical references.
ISBN 1-55111-471-2

  1. Language and logic.   2. Semantics (Philosophy).   3. Mathematics—Philosophy.
4. Frege, Gottlob, 1848–1925—Contributions in logic.   5. Russell, Bertrand, 1872–1970—
Contributions in logic.   I. Frege, Gottlob, 1848–1925.   II. Russell, Bertrand, 1872–1970.
III. Title.

B3245.F24S76 2003   160   C2002-904905-9

Broadview Press Ltd. is an independent, international publishing house, incorporated in 1985. Broadview believes in shared ownership, both with its employees and with the general public; since the year 2000 Broadview shares have traded publicly on the Toronto Venture Exchange under the symbol BDP.

We welcome comments and suggestions regarding any aspect of our publications –
please feel free to contact us at the addresses below or at broadview@broadviewpress.com.

North America
Post Office Box 1243, Peterborough, Ontario, Canada K9J 7H5
3576 California Road, Orchard Park, NY, USA 14127
Tel: (705) 743-8990; Fax: (705) 743-8353;
e-mail: customerservice@broadviewpress.com

UK, Ireland, and continental Europe
Thomas Lyster Ltd., Units 3 & 4a, Old Boundary Way,
Burscough Rd, Ormskirk, Lancashire L39 2YW
Tel: (1695) 575112; Fax: (1695) 570120
email: books@tlyster.co.uk

Australia and New Zealand
UNIREPS, University of New South Wales
Sydney, NSW, 2052
Tel: 61 2 9664 0999; Fax: 61 2 9664 5420
email: info.press@unsw.edu.au

www.broadviewpress.com

Broadview Press Ltd. gratefully acknowledges the financial support of the Government of Canada through the Book Publishing Industry Development Program for our publishing activities.

This book is printed on acid-free paper containing 30% post-consumer fibre.

Typesetting and assembly: True to Type Inc., Mississauga, Canada.

PRINTED IN CANADA

***Dedicated
to my father,
teacher and friend***

Contents

Preface ♦ 9
Notes on the Selections ♦ 11

Introduction ♦ 15
    I. Overview ♦ 16
    II. Philosophical Background ♦ 19
        i. Prevalent Ideas about Semantics in Modern Philosophy ♦ 19
        ii. A Sketch of Kant's Philosophy ♦ 21
        iii. Three Strands in Nineteenth-Century Philosophy ♦ 29
    III. The Logicist Thesis ♦ 31
        i. The Foundations of Mathematics ♦ 31
        ii. Frege's Program ♦ 37
        iii. Russell's Rediscovery of Logicism ♦ 42
    IV. Philosophical Logic ♦ 45
        i. Modern Logic ♦ 45
        ii. Function and Argument ♦ 54
        iii. Analysis of Language as Philosophical Method ♦ 60
    V. Some Disputed Issues in Early Analytic Philosophy ♦ 67
        i. Meaning: Semantic Monism and Semantic Dualism ♦ 67
        ii. Logical Form: On Denoting ♦ 75
        iii. Philosophy of Logic ♦ 82
    VI. The Legacy of Logicism ♦ 87
    Works Cited ♦ 89

Works of Gottlob Frege ♦ 91
    1. *Conceptual Notation* (1879), Preface and Chapter 1 ♦ 93
    2. On the Scientific Justification of a Conceptual
        Notation (1882) ♦ 119
    3. On the Aim of the "Conceptual Notation" (1882) ♦ 125
    4. *The Foundations of Arithmetic* (1884), Introduction ♦ 135
    5. Function and Concept (1891) ♦ 143
    6. On Concept and Object (1892) ♦ 163
    7. On Sense and Reference (1892) ♦ 175

8. What is a Function? (1904) ◆ 193
  9. The Thought: A Logical Inquiry (1918) ◆ 201

Works of Bertrand Russell ◆ 219
  10. Mathematics and the Metaphysicians (1901) ◆ 221
  11. On Denoting (1905) ◆ 235
  12. Knowledge by Acquaintance and Knowledge
      by Description (1911) ◆ 249
  13. Logic as the Essence of Philosophy (1914) ◆ 263
  14. Descriptions (1919) ◆ 279
  15. Mathematics and Logic (1919) ◆ 289

Sources ◆ 297

Preface

This collection began its life early in 2000, when I was putting together readings for a course on the origins of analytic philosophy, which I was preparing to teach for the first time. I found myself having to track down several different volumes and to copy several papers from each, and it occurred to me that an awful lot of philosophy instructors were doing exactly the same thing, for more or less the same course. I approached Broadview with the idea of putting together an anthology, and away we went.

I have discussed this project with several people over the last couple of years. Almost everyone suggests that it would be desirable to include some other material, in addition to works by Frege and Russell, in such a collection. The problem is that I have found nothing remotely approaching consensus as to precisely what else ought to be included. Wittgenstein's *Tractatus* is an obvious choice to complement these readings in such a course. But what else? Ramsay? Carnap? Tarski? Quine? A strong case can be made for all four of these, as indeed for some others. No small handful of philosophers or works stand out as the correct choices, as the consensus picks. It soon became clear that to try to do more than this would be to do a poor and patchy job of covering a wide and diverse terrain; I chose to aim for a good job of covering a smaller area.

Thus, an anthology of works by Frege and Russell, on logic, the philosophy of logic, and the philosophy of language. This volume collects together material that I consider necessary for an historical introduction to analytic philosophy or to the philosophy of language. For a one-semester course, it could be supplemented by Russell's "The Philosophy of Logical Atomism," Wittgenstein's *Tractatus*, and as much philosophy of mathematics as the instructor sees fit. (I have not given up on the idea of a second volume on these issues, which would cover subsequent important developments in the philosophy of language.)

The Introduction is unique in that it is framed around the questions: How did mathematical logic become such a principal instrument in the study of language? What do formal mathematical notions have to do with the study of meaning? These questions first occurred to me as an undergraduate, ten or twelve years ago. I think they are good questions, and that surprisingly little has been written that is explicitly about them. It is not that there are no good exegetical works on the history and substance of the developments at the origins of analytic philosophy—

indeed, my debt to fine works by Urmson (1956), Coffa (1991), and Kenny (1995), among others, may be evident at various stages of the Introduction. It is rather that expositors tend to presuppose rather than explain the evident connections between mathematics, logic, and language. So, to a large extent, the original scholarly and pedagogical contribution made in this Introduction consists in a careful explanation of how and why it is that questions about the foundations of mathematics, the development of modern logic, and the issues that make up the philosophy of language became so intimately linked during the late-nineteenth and early-twentieth centuries.

Queen's University funded much of the research necessary for the preparation of this volume, in the form of stipends from the Fund for Scholarly Research and the Advisory Research Council. During the latter stages of its preparation, I was funded by a Social Sciences & Humanities Research Council of Canada Postdoctoral Fellowship. I gratefully acknowledge this support.

Thanks to teachers, students, and friends for many rewarding discussions, and to the courteous and competent staff at Broadview. My deepest debt is to Adèle Mercier and Henry Laycock, for what they have taught me about these issues, and for continual reminders about how much there is to learn.

Notes on the Selections

[1] Preface and Chapter 1, *Conceptual Notation* (1879)

Strong cases can be made in support of the claims that this is the first work in modern logic, in the philosophy of language, and in the analytic tradition. The key innovation here is the function-argument analysis of language. It underlies the development of the propositional calculus, the predicate calculus, and the logic of quantifiers. The latter underlies the claim that logical form is in some sense hidden or obscured by natural language, which spurs the development of the analysis of language as a philosophical method.

Frege's notation is daunting, and some of his explanations wanting, but some of this is raw brilliance. First-hand acquaintance with it is a valuable thing. The Introduction motivates and orients the project, and Chapter 1 specifies the syntax and the semantics of the language. (There are two subsequent Chapters, not reprinted here. In Chapter 2 Frege sets out to derive laws of logic from the fewest possible axioms; in Chapter 3 he defines some technically important relations.) A note on terminology: Frege's title is 'Begriffsshrift', which translates literally as 'concept script;' other translations for this term include 'logically perfect language' and 'ideal language.' Since this volume includes Bynum's (1972) translation, I shall follow his usage of 'conceptual notation.'

[2] On the Scientific Justification of a Conceptual Notation (1882)

This paper is, first and foremost, a plea for people to read [1]. It contains some helpful explanation of the project, and an argument, premised on the logical defects in natural languages, for why an artificial language is required for the study of logic.

[3] On the Aim of the Conceptual Notation (1882)

Some of the few reviewers of [1] dismissed it as a poor substitute for Boole's (1854) *Laws of Thought*. Frege explains the differences between his project and Boole's, and some of the advantages of his system over Boole's. This paper contains much further insight into the substance of [1].

[4] Introduction, *The Foundations of Arithmetic* (1884)

This book is the second step toward establishing the logicist thesis, the first step being the system of deduction developed in [1]. The project here is a logical definition of the concept of number. This short excerpt is included to give the reader a sense of the method and the content of the logicist project. Also, the three principles outlined at the end are seminal in the philosophy of language.

[5] Function and Concept (1891)

I refer to [5]-[8] as 'Frege's middle papers.' They largely consist of elaborations of the beauty and power of the function-argument analysis of language. In these papers Frege explains how and why this approach applies beyond the confines of logic and mathematics. This kind of analysis eventually leads to a new way of doing philosophy, as translating propositions into the function-argument calculus is shown to solve or dissolve some puzzles and problems.

[5] and [8] complement each other nicely: given the proper understanding of what constitutes a function, explained in the early pages of [5] and more thoroughly in [8], [5] explains how the function-argument distinction offers an elegant explanation of many semantic phenomena. Treating linguistic expressions as functions is not only the key to a better notation for logic; further, the notion of a function, properly understood, captures the essence of predication, and applies to many other sub-sentential phenomena.

[6] Concept and Object (1892)

A response to Kerry's review of *Foundations of Arithmetic*, in which he accuses Frege of misusing the term 'concept.' This paper affords a unique look into some corners of Frege's semantic system.

[7] On Sense and Reference (1892)

The sense-reference distinction is another key development in the creation of the philosophy of language as an autonomous area of inquiry. Philosophers have yet to understand completely the upshot of the considerations Frege raises in this paper.

[8] What is a Function? (1904)

This paper consists of criticisms of some mistaken ideas about the semantics of functions. Frege shows that many of his contemporaries confuse functions with their symbolic expression and with their values. The function-argument analysis depends on avoiding such semantic sloppiness: the claim that many linguistic expressions behave semantically like functions lacks clear bite until we are precise about what a function is.

[9] The Thought: A Logical Inquiry (1918)

This is the first chapter of a book on logic that Frege worked on intermittently, without ever finishing. It contains some of the clearest statements of his views on the nature of logic and language.

[10] Mathematics and the Mathematicians (1901)

This is a good readable survey of many of the things one needs to know about nineteenth-century mathematics in order to understand these philosophical developments. I assign this as the first reading of the course, as part of the necessary background to [1].

[11] On Denoting (1905)

This is the first statement of the theory of descriptions. The arguments for the theory are both brilliant and infuriating. Among the many notable features of this classic is that it is one of the very few places where Russell offers an argument directed at Frege's approach to semantics.

[12] Knowledge by Acquaintance and Knowledge by Description (1911)

This paper is an elaboration of the epistemological consequence of the theory of descriptions. Russell took the discovery of knowledge by description to be among his most important contributions to philosophy.

[13] Logic as the Essence of Philosophy (1914)

This is the second chapter from *Our Knowledge of the External World*. It is a manifesto for philosophical logic, a manual explaining how (and why) to be an analytic philosopher. Noteworthy features include detailed explanations of how philosophers have been led astray by inattention to semantics, and musings about the epistemological upshot of some recent developments in logic.

[14] Descriptions (1919)

This is the most accessible statement of the key features of the theory of descriptions. I assign this in conjunction with [11]. It is the sixteenth chapter from *Introduction to Mathematical Philosophy*.

[15] Mathematics and Logic (1919)

This is the eighteenth and final chapter from *Introduction to Mathematical Philosophy*. It offers a nice clear statement of Russell's logicist thesis, as well as an exploration of some deep and difficult questions in the philosophy of logic.

Introduction

The purpose of this Introduction is to help the reader understand the contributions made by Gottlob Frege (1848-1925) and Bertrand Russell (1872-1970) to the methodology and subject matter of twentieth-century philosophy. It is focused on the question: How did mathematical logic become such a principal tool in the study of language? The answer to that question constitutes a seminal chapter in the history of ideas, one that repays careful study. Toward the end of establishing the thesis of logicism—i.e., the thesis that mathematics is a branch of logic, in the sense that the truths of mathematics rest only on the laws of logic—Frege and Russell lay the groundwork for the development of modern logic and for the creation of the philosophy of language as an autonomous area of inquiry. In so doing, they also develop an analytic approach to philosophical questions, which has had enduring and pervasive effects on the argumentative methods and the standards of rigor that characterize the discipline of philosophy.

The structure of the plot is as follows: Immediately below I introduce some key terms and sketch some of the relevant terrain. Part 2 describes the intellectual climate in which Frege and Russell were educated, with specific attention to prevalent ideas about meaning. In Part 3 I identify some problems in logic and the foundations of mathematics that exercised many thinkers in the nineteenth century, and describe Frege's and Russell's logicist programs of using logic to cement the foundations of mathematics. Part 4 explains the tools and methods of the project of philosophical logic. Part 5 is a discussion of some rather fundamental issues about which Frege and Russell disagreed, and which form the basis of some subsequently productive areas in the philosophy of language and logic. Finally, Part 6 is a brief overview of the enduring impact of these works by Frege and Russell.

These works by Frege and Russell are exemplary instances of the maxim that semantic analysis is a necessary precursor to careful systematic inquiry. This maxim instructs us to pay careful attention to such concepts as meaning, definition, truth, and inference before we can productively argue about the merits of philosophical theories or views. My principal aim here is to explain the genesis and substance of this influential approach to philosophical problems. (References in square brackets refer to the works included in this collection. The rest of the works cited are collected at the end of this Introduction.)

I. Overview

Among other things, these essays constitute an anthology on the concept of *logical form*, as the concept was conceived by two of its first and most insightful students. So I must describe briefly, and situate in historical context, the concept of logical form. (Warning: things get a little complicated, here. I see no way around beginning with some difficult concepts. These concepts must be first sketched in a preliminary way, and our grasp of them will be fleshed in, will grow more comprehensive, as the story proceeds.)

Utterance will be used to denote a vocalization or inscription of a linguistic expression for which the question of truth or falsity arises. So, for instance, 'is' does not count as an utterance, but 'John is dancing' does, because there is no question about the truth or falsity of 'is.' (It is a bit forced to call the act of writing a sentence the making of an utterance, but when the alternative is to repeat the cumbersome 'utterance or inscription of a linguistic expression for which the question of truth or falsity arises,' it is worth putting up with such awkwardness.) *Proposition* denotes the meaning or content expressed with an utterance. As a first approximation, we could view a proposition as built from the two components of (i) the meanings of the constituent expressions and (ii) a certain logical form. The logical form is the underlying structure of the proposition, which it can share with distinct propositions that differ in meaning. It is a sentence-schema, devoid of any particular content, which gives the structure of the relations among the bits that compose the proposition. Consider the following:

(1) Calgary is larger than Winnipeg.
(2) Sue is taller than Tim.
(3) Every politician is corrupt.
(4) Whales are mammals.

(1) and (2) have the same logical form, which could be represented as 'xRy'—they assert that the same type of relation holds between the referent of one proper name and the referent of another. Likewise (3) and (4)—both assert that to anything to which a certain predicate applies, a second predicate also applies; so the logical form they share could be represented as 'All Fs are Gs.'

Why call this notion *logical* form? The answer to that question depends on the meaning of 'logic.' Here is Frege's characterization of logic:

> ...to make a judgement because we are cognizant of other truths as providing a justification for it is known as inferring. There are laws governing this kind of justification, and to set up these laws of valid inference is the goal of logic. (Frege 1979: 3)

Logic is the study of inference, of what sets of premises support or justify in virtue of the relations among their parts, and irrespective of their particular meaning or content. Patterns of inference that are structurally sound, or good by virtue of the formal relations among their parts, logicians call 'valid.' Consider the following:

(5) Humans are mammals, and mammals are mortal. Therefore, humans are mortal.

(6) If a number is even, then it is divisible by 2. 6 is even. Therefore, 6 is divisible by 2.

What makes (5) and (6) good inferences has nothing to do with the particular meanings of the specific terms involved. They are good inferences solely in virtue of the relations among their parts. In other words, (5) and (6) are instances of the valid logical forms:

(5*) All Fs are Gs, All Gs are Hs. Therefore, all Fs are Hs.

(6*) All Fs are Gs, x is F. Therefore, x is G.

All arguments of these forms are good ones, are structurally sound. Consider, for instance:

(5a) Politicians are corrupt, and corrupt people are dishonest. Therefore, politicians are dishonest.

(6a) Politicians are corrupt, and Fred is a politician. Therefore, Fred is corrupt.

(5) and (5a), and (6) and (6a), are substitution instances of the same logical forms: they are structurally identical, and differ only in the interchange of one set of terms for another. They have the same logical scaffolding—only their matter differs—and are valid for exactly the same reasons. These reasons are the province of the logician. Logic is the study of these general, repeatable, structural properties of inferences, of these logical forms.

It is important to note that not all good inferences have this structural property of validity; and so not all good inferences are of interest to logicians. Consider, for instance:

(7) This sweater is blue. Therefore, it is not red.

(7a) 6 is even. Therefore, 6 is not odd.

These are perfectly good inferences—the conclusion is fully supported by the premise. However, the goodness of these inferences falls beyond the general purview of logic, because it has to do with the particular meanings involved, and not with the way in which they are arranged or structured. The logical form underlying these inferences is:

(7*) x is F. Therefore, x is not G.

There are many bad inferences of this form, such as:

(7b) Mary is human. Therefore, Mary is not Canadian.

(7c) 2 is even. Therefore, 2 is not prime.

Thus, even though (7) and (7a) are good inferences, their goodness is not a matter of logic. (7*) is an invalid logical form.

Logic is solely the study of formal or structural entailment relations. If it were also up to logic to tell us which inferences are good ones in virtue of the meanings of the terms involved—i.e., for each concept F, which other concepts are compatible with it, which are inconsistent with it, and so on—then logic would pretty much be the study of everything. In contrast, logicians are experts about a certain kind of inference: they are interested only in the inferential properties that utterances with the same logical form have in common. So, these sentence-schemas—'xRy,' 'x is F,' 'All Fs are Gs' and so on—are called *logical* forms because they capture all and only that which is relevant to the formal or structural relations of entailment, all and only that which is pertinent for the study of logic.

Our conceptions of the notions of both logical form and meaning will be refined throughout the course of this Introduction, particularly in Parts III through V.

It is not too much of a stretch to say that traditional logic recognized only one logical form—i.e., 'S is P,' where 'S' stands for the grammatical subject and 'P' for the grammatical predicate. (This is a bit of a stretch, though—'all,' 'no,' or 'some' must precede the subject-term 'S,' and 'not' can precede the predicate-term 'P.' However, such complications do not undermine these introductory points.) The importance of the concept of logical form comes with the recognition that forcing all utterances to fit this all-purpose 'S is P' mold is not sensitive to some important semantic distinctions, and will result in licensing some shoddy inferences. Frege and Russell are among the first to identify and diagnose the problems that result from putting all grammatical subjects into the same logical category, and from likewise treating all grammatical predicates as logically on a par. They allege that while 'S is P' does reflect the essential grammatical parts of a sentence, it does not capture that which is relevant for the study of logic. This allegation will be substantiated in detail in Part IV.

The recognition that there are diverse varieties of logical form constituted one of the key points of departure of modern from traditional logic, and spurred a closely related project of semantic analysis called *philosophical logic*. Most generally, philosophical logic is the study of the properties of, and relations between, propositions, of how the meanings and logical forms of the expressions uttered, as well as features of the context of utterance, determine the content of the propositions expressed. Although not without some historical precedent—most notably, Aristotle's *Organon* and the medieval tradition it spurred—modern philosophers tended to be insensitive to the issues and questions of philosophical logic. (The modern period in philosophy runs, roughly, from the Enlightenment through to the nineteenth century. Modern logic, in contrast, begins in the late nineteenth century. The awkward truth is that the development of modern logic was one of the factors that ushered in the end of modern philosophy. These developments will be sorted out in some depth in Parts II and III.)

Slogans for this project of philosophical logic include Russell's (1918: 269) allegation that "practically all traditional metaphysics is filled with mistakes due to bad grammar," and Wittgenstein's (1961: 106) "Distrust of grammar is a first requisite for philosophizing." In this vein, Russell ([13]: 270) contains several tirades against the "many errors in traditional metaphysics" which result from the "failure to perceive" that, from a logical point of view, several kinds of utterance are not of the subject-predicate form. All of the writings in this collection are peppered with formulations of this sentiment. A central task of philosophical logic is to systematically identify the diverse varieties of logical form, and to study their various roles in philosophical argument; its avowed enemy is a blindness to fine semantic distinctions, and a tendency to force the diverse varieties of logical form into the one traditional 'S is P' mold.

Unlike formal logic, philosophical logic is not restricted to formal or structural patterns of inference. The province of philosophical logic also includes the study of the meaning of concepts and propositions, and of content-specific patterns of inference. (The first great work in this tradition of philosophical logic is Frege's (1884) *Foundations of Arithmetic*, subtitled a "logico-mathematical inquiry into the concept of number"; Russell's first sustained contribution comes with his (1903) *Principles of Mathematics*. In Part III I explore why mathematics is the area in which philosophical logic first takes root.) It is in the course of this project that the tools and methods of modern logic spill over the boundaries of logic and mathematics, and are applied more broadly to all manner of philosophical problems. In this way, the developments catalogued in this volume constitute a high-water mark of the ancient methodological doctrine that the analysis of language is an important philosophical method.

The need for this kind of logical and semantic work became evident to both Frege and Russell in their attempts to tidy up the foundations of mathematics. Indeed, at least initially, their interest in logic is largely instrumental, working toward the end of establishing the logicist thesis that mathematics is a branch of logic. By and large, that thesis was ill-fated, but their contributions to the study of logic, of thought, and of language made in the course of its defense are monumental. There are several reasons why the notion of logical form belongs at the base of that monument. To underline three: the tools and methods Frege and Russell introduce in the course of their studies of logical form are the basis of their considerable contributions to the development of modern logic; a new and productive analytical method in philosophy grows out of these studies; and the efforts and successes of Frege and Russell in the study of logical form establish and set the agenda for the philosophy of language as an autonomous area of inquiry. Their work on the concept of logical form spurs a period of unprecedented semantic and logical work, and it has affected all corners of subsequent philosophical inquiry.

II. Philosophical Background

II.i Prevalent Ideas about Semantics in Modern Philosophy

What I classify as modern semantic ideas are views about meaningfulness and communication that were widely held at least from Descartes through Kant and well into the nineteenth century. These views were rarely made explicit—notable exceptions include Arnauld and Nicole's (1662) *Port Royal Logic* and Locke's (1690) *Essay Concerning Human Understanding*—which is not surprising, given that semantic questions were so rarely addressed explicitly during this period. Rather, when you scratch the surface of what various philosophers have to say, when they brush up against the ubiquitous topics of meaning and the nature of language, something of a consensus can be found. (Beware a consensus among philosophers: it often betrays unquestioned assumptions held in common, rather than conclusions separately and disinterestedly derived.)

The crux of this prevalent modern semantic package is that the meaning of an expression is constituted by the ideas, images or impressions in the head of the speaker. Arnauld and Nicole, for instance, define a word as a sound which is used to express an idea (Part II, Ch. 1); Locke claims that words "stand for nothing but ideas in the head of him that uses them" (Book III, ii, 2). This affords quite a simple picture of meaningfulness: a linguistic expression signifies an idea in the head of the speaker, and that idea represents a referent. Two tenets can be seen to underlie the view: [1] first-person authority about meaning (i.e., the speaker is autonomous as to the meaning of the words uttered) and [2] the doctrine that meaning is constituted by subjective psychological content (i.e., the meaning of an expression is a mental image or idea which the speaker has in mind in making the utterance).

The limitations of this private-psychological approach to meaningfulness are well known, largely due to Frege and Wittgenstein. Perhaps the most damning among its many problems are that there are a wide range of meaningful linguistic expressions—from 'of' to 'game' to 'politics'—which do not signify any idea or image in particular, and that ideas are, but linguistic meaning is not, the private property of individual speakers. Frege argues tirelessly against the notion that linguistic meaning could be constructed from anything subjective; Wittgenstein raises a host of considerations against the tenet of first-person authority. It is now generally acknowledged that meaning cannot be identified with, or constructed from, private psychological content: we use language to talk about things in the world, not just about our ideas of them. Neither tenet of this modern semantic package enjoys much currency even individually, let alone when conjoined.

There are two reasons why I start this Introduction here. The first is that an integral part of understanding any philosophical movement is a grasp of what it is a reaction against. Most of Frege's and Russell's predecessors and contemporaries belong in this modern semantic tradition; to some extent, their semantic and logical explorations, on the path to proving the logicist thesis, are spurred by dissatisfaction with this approach to meaning. The second reason is more technical. It has to do with the danger of an act-object ambiguity that is inherent in this approach to semantics. Terms that refer to psychological states or processes, such as 'judgement' or 'representation,' admit of such ambiguities. For instance, the expression 'the judgement that 2+2=4' is ambiguous—'judgement' here might denote the mental act of judging, or might denote that which is judged, the object at which that mental act is directed. The former, the act, is something that people do when adding figures; the latter, the object, is a truth of mathematics. Similarly, 'representation' is sometimes used to denote the act of representing and sometimes used to denote the object of that act, that which is represented. The former are things that people do (for instance, to conjure up a pretty image of Vienna); the latter are things that people entertain or express (for instance, the proposition that Vienna is pretty). Only the former, the act, occurs entirely within the head of someone in particular. The latter, the object of the mental act—the propositions that 2+2=4 and that Vienna is pretty—can be expressed and entertained by any number of people.

So, clearly, such psychological terms can denote both psychological acts and their object, and it is crucial to distinguish these things if we are to be precise. However, this is difficult to do, on the modern strategy of constructing linguistic meaning from subjective psychological content. That approach engenders a tendency to conflate the psychological acts of expressing or grasping something meaningful with the content of the object of those psychological acts, with the proposition that is grasped or expressed. This approach to meaning does not encourage, and perhaps does not even allow for, a clear act-object distinction. This point will be revisited at several junctures in the following pages, because many of the questions which set philosophical logic and the philosophy of language rolling can only be formulated once psychological acts are carefully distinguished from the meaningful objects at which thoughts and utterances are directed.

In this respect, Frege is a pioneer. He constantly warns against the dangers attendant upon not paying sufficient heed to the difference between subjective mental items or processes and the objects at which such psychological acts are directed. (Cf., e.g., ([4]: 138-40).) Russell, too, makes much of this "distinction between act and object in our apprehending of things" (Russell 1912: 42). This distinction is crucial to the creation of the philosophy of language as an autonomous area of inquiry, because to draw this distinction is to reject the notion that the study of meaning is a sub-discipline within psychology, and hence forces the need for other ways and means of studying the intersubjectively accessible content of our thought and talk. A given proposition may be entertained by any number of mental acts, and may be expressed with quite different utterances, in different contexts, by different speakers, and in different languages. This raises one of the central metaphysical questions within the philosophy of language: if propositions are distinct from both particular mental acts and particular linguistic expressions, then what exactly are they? This question will be discussed in some depth in section V.i.

II.ii A Sketch of Kant's Philosophy

Kant's thoroughly original approach to understanding the relation between mind and world has had profound and enduring effects on the confines and substance of many perennial philosophical debates. His legacy was the dominant force in Western philosophy throughout the nineteenth century and well beyond—an awful lot of subsequent work is explicitly situated in relation to Kant. The work of Frege and Russell is no different in this respect, so it is important here to explain some features of the Kantian climate in which they conceived and constructed their work. What follows is far too brief to serve as an introduction to Kant's philosophy; and paraphrase inevitably involves some degree of butchery of Kant's marvelously coherent and complex system. My aim here is just to underline some important points of Kant's philosophy to which Frege or Russell respond. Specifically, I have three goals: [1] to unpack the building blocks of Kant's semantics, [2] to discuss the notion of a synthetic *a priori* judgement and its role in Kant's philosophy, and [3] to outline Kant's views on mathematical and logical truths. (All references in square brackets are to the Kemp-Smith translation of the *Critique of Pure Reason*.)

[1] *The building blocks of Kant's semantics*

(a) A *judgement* is the putting together in thought of a subject and a predicate. Judgements are the sorts of thing for which the question of truth arises—the predicate qualifies the subject in some way (e.g., 'Gold is valuable,' 'Whales are mammals'), and the judgement is true if and only if the predicate does indeed fit or inhere in the subject-term. (Note the act-object ambiguity here, with this term 'judgement.')

(b) *Analytic/synthetic* is a distinction between two types of judgement that is absolutely central to Kant's philosophy. Kant uses the principle of contradiction and the metaphor of containment to explain the notion of analyticity: to deny the truth of an analytic judgement is to make a contradictory judgement, one which is false in virtue of the meaning of the constituent terms; an analytic judgement is one in which the predicate is contained in the subject. Examples Kant gives of analytic judgements include 'Gold is a yellow metal' and 'All bodies are extended.' To claim that these judgements are analytic is to say that they afford no new information to someone who already understood the subject-concept, and that judgements such as 'This is made of gold, but it contains no metal' or 'This body is not extended' are self-contradictory. In a synthetic judgement, in contrast, the predicate is not already contained in the subject. Synthetic judgements do constitute substantive increases in our knowledge of the subject-concept, and one does not contradict oneself in denying the truth of a synthetic judgement. Examples of synthetic judgements include 'Gold is valuable' and 'All bodies are heavy.'

Analysis, for Kant, is the process of breaking down a complex concept into its constituents. The end of analysis is a complete, exhaustive definition of the subject-concept. Significantly, Kant holds that analysis is not a means of acquiring new knowledge, that analysis cannot really tell you anything that you did not already know. The logicists reject this view of analysis, holding that philosophical analysis can significantly build on and amplify what one already knew. (Frege (1884: 101) employs a germane metaphor to illustrate this view: there is a sense in which the content of an analytic judgement is contained in the subject-concept, "but as plants are contained in their seeds, not as beams are contained in a house.")

(One problem with Kant's distinction is the lack of a firm criterion for distinguishing analytic from synthetic truth. Suppose, for instance, that two people disagree as to whether 'Gold is a yellow metal' should be classified as analytic or synthetic. The metaphor of containment is of little help here, as is the principle of contradiction. They merely provide means of reformulating the question: our disputants disagree as to whether 'gold' contains 'yellow metal,' as to whether the concept of non-yellow gold is self-contradictory; and there is nothing in Kant's philosophy by means of which to resolve this dispute. From the logicists' vantage point, Kant's distinction between analytic and synthetic is a paradigm case of a weighty philosophical doctrine resting on a shaky semantic foundation. Because of the role that this distinction plays in his sprawling philosophical project, this inattention to semantic matters at the initial stages of inquiry threatens the integrity of the whole edifice.)

(c) *A priori/a posteriori* is a second distinction that is crucial to the Kantian taxonomy of judgements. It is an epistemological distinction, concerning what it takes to know that a judgement is true. An *a priori* judgement is one that can be known to be true without having experienced that which the judgement is about. All analytic judgements are knowable *a priori*: Kant holds that necessity and universality constitute a sufficient condition for apriority—i.e., if a judgement is necessarily and universally true, then it is apriori—and, with a nod to Hume, he holds that knowledge of the truth of a necessary universal judgement cannot be grounded in experience. Any judgement that is universal and necessary (for instance, 'All triangles are three sided,' or 'Every event has a cause') cannot be justified experientially; hence any such judgement that is knowable must be knowable *a priori*.

In contrast, if knowledge of the truth of a judgement does depend on experience of what the judgement is about, then the judgement is knowable only *a posteriori*. *A posteriori* judgements (such as 'Canada is larger than Brazil,' 'It rains more in Sao Paolo than it does in Montreal') are neither universal nor necessary—they are about states of affairs that may or may not have actually obtained. All *a posteriori* judgements are synthetic, according to Kant: wherever experience is required to verify a judgement, its subject-concept is not already contained in the predicate.

So all analytic judgements are knowable *a priori*, and all *a posteriori* judgements are synthetic; but neither of these entailments is reciprocal, or bi-conditional.

(d) A *representation* is the most basic semantic unit for Kant, the bits out of which judgements are composed. Squarely within the modern semantic tradition, Kant holds that representations are "modifications of the mind" which "belong to inner sense" [A99], psychological entities which constitute the meaning of the terms which compose judgements. 'Intuition' is the faculty that affords the content of representations. Kant's maxim that concepts without intuitions are empty while intuitions without concepts are blind suggests that intuition supplies raw matter and that concepts are the form imposed on that matter in significant thought.

Typically, the content of a representation is supplied via the senses—if you want to know what 'pain' or 'red' or 'sweet' means, get out in the world and start experiencing. Kant holds that most of the meanings that are known to, or graspable by, a subject have as their content representations afforded by this empirical sort of intuition. However—and this is one lesson learned from Hume—empirical intuition cannot provide all the representations required to account for what Kant takes to be our actual knowledge. Another sort of intuition is needed to supply the constituent bits of our judgements that transcend the confines of our particular experiences—to supply our universal and necessary knowledge. Kant names this sort of intuition 'pure intuition.' We know that every event has a cause; empirical intuition cannot supply the content of that judgement, because its range—i.e., *every* event—extends beyond the bounds of our experience; and hence the faculty of pure intuition is posited to explain our grasp of the content of such non-empirical judgements.

Much has been written on Kant's move here, about whether this appeal to pure intuition is legitimate and whether it can do the work that Kant demands of it. We are on the verge of some complicated questions of Kantian exegesis. I will go no further into these questions here, though; what follows is a general overview of what Kant attempts to construct from these building blocks.

[2] *Synthetic* a priori *judgements and their role in Kant's philosophy*

We have knowledge of judgements whose content goes beyond the bounds of our experience, and therefore there is a non-empirical source of intuition. How is this possible? Given that our knowledge can extend beyond the bounds of what we have experienced, what can we deduce about the mind?

Kant's seminal answer to these questions involves a radical hypothesis about the relation between mind and world. At [Bxvii], Kant uses the following analogy to explain the hypothesis: astronomers prior to Copernicus had assumed that they are stationary and that the stars revolve around them; Copernicus tried the hypothesis that observers on earth are in motion and the stars are, relatively, at rest, and the result was a great leap forward in our understanding of the universe. In the analogy, the empiricism which Hume had followed to its skeptical end—more specifically, the notions that the mind is a passive recipient of information, and that all knowledge consists of sensory input and inductions therefrom—is compared to pre-Copernican astronomy. Kant's innovative Copernican hypothesis is the idea that the mind plays an active role in synthesizing and categorizing sensory input, that the mind is the director, as opposed to the audience, in conscious experience.

Hume had demonstrated the impossibility of grounding in experiential input the consistency and regularity of the world. Kant's hypothesis is that, to some extent, the mind constructs and constitutes this consistency and regularity. Kant provides an elaborate and marvelously comprehensive theory of a non-empirical source of knowledge: there are active faculties of mind, constitutive powers of mind, which categorize and give form to the matter supplied to mind via the senses, and thereby structure and constitute our knowledge.

The cornerstone of Kant's philosophical edifice is the synthetic *a priori* judgement. According to Kant, synthetic *a priori* judgements are logically possible in that the concept of a synthetic judgement does not entail or contain the concept '*a posteriori*,' and neither does '*a priori*' entail or contain 'analytic.' Synthetic *a priori* judgements are possible for agents like us, given the Copernican hypothesis that the mind is an active synthesizer of sensory input. Kant's (1784) *Critique of Pure Reason* is dedicated to establishing the legitimacy of these synthetic *a priori* judgements, on the grounds that this hypothesis does some-thing that none of the alternatives can: namely, account for our actual knowledge. Given the hypothesis that some of our judgements are contributions to, rather than inductions from, experience, there is conceptual space for judgements which are substantial extensions of human knowledge, unlike trivial analytic judgements, but which are not given in, or generalizations from, experience.

According to Kant, as in the case of Copernicus' Revolution, what we have here is a huge step forward in human knowledge. As the constitutive powers of mind ground synthetic *a priori* judgements, synthetic *a priori* judgements ground philosophy itself. For the first time, we have an explanation of how pure reason—*a priori* thinking—can increase the volume of human knowledge. Prior to Kant, there is analysis, an *a priori* activity that cannot really tell us anything we did not already know, and there is observation and induction. Hume showed us how little knowledge can be secured by those faculties, unaided. What Kant has discovered is a type of *a priori* knowledge, which is thus universal and necessary, but which yet has real empirical bite, and can constitute significant extensions of human knowledge. For Kant, all significant scientific advances (such as the principle of the conservation of matter) and all mathematical truths (such as that captured by the formula '$a^2+b^2 = c^2$') are synthetic *a priori* judgements—they go well beyond experiential input, and they add to our knowledge of the subject-concept. All significant philosophy consists in the discovery and elucidation of synthetic *a priori* judgements, judgements such as 'Every event has a cause' and 'Humans are free agents.' As opposed to the views of the logicists and of their successors in the analytic tradition, and as is perhaps most clearly embodied in the systems of Kant's successors, Hegel and Bradley, all significant philosophy is synthesis, not analysis, on a broadly Kantian approach. Given the possibility of these universal necessary judgements that constitute genuine extensions of the body of human knowledge, their importance to philosophy or to human knowledge generally cannot be overestimated.

[3] *Kant's views on mathematical and logical truth*

The private psychological approach to linguistic meaning outlined in section II.i fits quite well with a view about the nature of logic and mathematics that prevails among empiricists and idealists. The view is called 'psychologism'; the term applies to any theory according to which logic or mathematics are in some sense branches of psychology, disciplines whose aim is to study how humans actually reckon, as opposed to the study of how one ought to think, or of the objects and properties involved in calculation. Psychologism is Frege's most hated and enduring enemy—almost everything he wrote contains a stab at some form of psychologism, as it is opposed to his realism about logical and mathematical truth. ('Realism' in this context denotes the view that logical and mathematical truths are mind-independent and objective; that they are eternal truths that humans discover, as opposed to contingent regularities that we create or construct. Russell campaigns against psychologism as well, and is at most stages of his career a staunch realist about logic and mathematics as well. Cf. section V.iii for more on this point.) Kant's views on logic and mathematical truth are, at least implicitly, psychologistic. I will make one comment about Kant's views on logic, one comment about his views on mathematics, and then highlight some consequences of these points that will be significant in what follows.

The remark about logic can be centered on Kant's infamous ill-fated declaration that logic has not advanced a single step since Aristotle, and constitutes a closed and complete body of doctrine [Bviii]. The falsity of this statement will be attested in the following pages—much progress was made in the study of logic in the nineteenth century, and its growth since the work of Frege and Russell has been exponential. Further, this declaration betrays a deep divergence between Kant and the logicists on the nature of logic. Logic for Frege and Russell is the study of truth-preserving inference. On this view, logic could never be complete: there are an infinite number of patterns of inference to identify and catalogue. So the above claim shows that something other than the prescriptive study of inference is involved in logic, that there is an element of description in logic, for Kant.

Indeed, Kant's views on how the mind actually works are integrally tied up with the categorical principles of Aristotelian logic. Kant's categories, key elements of the constitutive powers of mind, are crafted to fit with the pillars of traditional logic. This holds true, more generally, of many of Kant's theses about how the actively synthesizing mind actually works. There is no rigid boundary in Kant's thought, as there clearly is in the works of Frege and Russell, between the study of logic and the empirical study of how humans actually think. (Note that this, *per se*, is no criticism of Kant's epistemology, or of his metaphysics: there are a host of plausible and interesting links between logical prescription and the description of rational processes. The point is noteworthy because of the divergence from the logicists it illustrates concerning the nature of logic. This psychologistic closeable and completeable project is not the logic to which Frege and Russell hoped to reduce mathematics.)

Divergences between Kant and the logicists are even more pronounced in the case of mathematics. Explicitly in the writings of both Frege and Russell, logicism itself is an effort to save mathematics from Kant. Kant holds that mathematical judgements are synthetic *a priori*, a view which he explains at [B15-17] with reference to the judgement that 7+5=12. Such judgements are knowable *a priori*, because they have the characteristic marks of universality and necessity. However, they are, according to Kant, not analytic: "The concept of twelve is by no means thought by merely thinking of the combination of seven and five; and, analyze this possible sum as we may, we shall not discover twelve in the concept." Here again we risk opening up some frightfully big cans of exegetical worms, with the questions: Are mathematical judgements synthetic? What are Kant's arguments for his view? Let me just explain how Kant's view becomes less plausible if we attend carefully to the act-object distinction.

There is a sense in which all judgements are synthetic: any judgement, qua mental act, requires a bringing together of, and synthesizing, subject and predicate. So it is to the content of the objects judged that the analytic-synthetic distinction significantly applies. However, Kant's reasons for classifying '7+5=12' as synthetic have primarily to do with the mental acts which constitute making or grasping the judgement—for instance, the act of thinking '7+5' is distinct from, and does not contain, the act of thinking '12;' since grasping the contents which flank the identity sign require different mental acts, this seems to be a case wherein something has been added to our subject-concept; and so on. If we attend to the distinction between the act of judging and the object judged, Kant's considerations do not constitute good reason to classify the content of the object judged as synthetic. Qua object, the proposition that 7+5=12 is a strong candidate for analyticity: it consists of the very same content, denoted in different ways, flanking the identity sign. Notwithstanding the synthetic work the mind might have to do as it grasps the truth expressed in the equation, there is a very clear sense in which one makes a contradictory judgement in sincerely and thoughtfully judging that $7+5 \neq 12$: namely, the judgement could not possibly be true. Such a judgement is false in virtue of the meanings of the terms involved, which constitutes good reason to classify the proposition expressed as analytic.

Logicism can be characterized as an attempt to establish the analyticity of arithmetical truth—no doubt, the most thorough and systematic such attempt ever undertaken. Toward that end, the logicists saw it necessary to embark upon comprehensive preliminary semantic and logical analyses. Let me highlight a couple of points, and then move on with that story. First, Kant takes all judgements to be of the subject-predicate form. (This, too is a pillar of traditional logic, as is instanced in the generally applicable logical form 'S is P'; its rejection is a point of departure for modern logic, and for the project of philosophical logic.) Frege and Russell identify counterexamples to and limits of this doctrine. For instance: What is the subject of '7+5=12'? Is '12=7+5' exactly the same judgement? Qua act, perhaps not; qua object, almost certainly. Regardless, one theme which will recur below (cf. especially section IV.i) is that not all judgements consist of the putting together of a subject-concept with a predicate-concept, and that forcing all statements into that mold can lead to all manner of erroneous inferences. For instance, Frege draws attention to the importance for mathematics of several kinds of judgement in which "there can simply be no question of a subject in Kant's sense" (1884: 100); and points out that "language has means of presenting now one, now another, part of [one and the same] thought as the subject" ([6]: 169). As noted above, the limitations and pitfalls inherent in the subject-predicate approach to all judgement is a principal theme running through Russell's writings.

Another important point which has come up already, and will recur in the following pages, is that Frege's and Russell's efforts to ground the analyticity of mathematical truth is undergirded by the view that analysis, rather than synthesis, is the philosopher's primary task. The idea that the analysis of language constitutes an important philosophical method is at work in virtually all of Plato's dialogues, and is endorsed by Aristotle; Frege, Russell, and their successors revive and develop this ancient tradition further. Again, note well that analysis for the logicists means something different from Kant's conception of analysis as the trivial unpacking of what we already knew. Frege derides "the widespread contempt for analytic judgements" and its attendant "legend of the sterility of pure logic" (1884: 24). His metaphor that the fruits of analysis "are contained in the definitions, but as plants are contained in their seeds, not as beams are contained in a house" (1884: 101) underlines the seminal role which analysis would play throughout twentieth-century philosophy.

One last point: there is a connection between the difference between the logicists and Kant on the philosophical worth of analysis and the difference between the logicists' realism and Kant's psychologism about mathematical and logical truth. Frege and Russell hold that mathematical objects and logical truths are mind-independent, while there is a sense in which mathematics and logic are all in the head for Kant. While the logicists conceive of logical and mathematical analysis as the study of objective entities, for Kant such projects are just unpacking the analytical presuppositions of psychological acts or objects. Hence, it is no accident that thinking through properties of, and relations between, mathematical objects or logical truths—even if they are contained in the premises from which one starts—is viewed as a more significant and more edifying endeavor by the logicists. It is not hard to see why there is a more substantive role for analysis to play, on the view that the objects of analysis are mind-independent. (This is not to say that there is any straightforward necessary connection, in either direction, between being a realist about logical or mathematical objects and the logicists' view of analysis; rather, I put this forward as a connection which is worth thinking about.)

The analytic movement, the basement of which we are about to enter, is, to a large extent, the study of semantic content, of the content of judgements, as distinct from psychological acts and ideas. Frege and Russell are united in opposition to some key assumptions at the base of Kant's philosophy. Symptoms include the lack of a firm criterion for analyticity and a doctrine of analyticity that applies only to the limited set of subject-predicate statements. The disease those symptoms betray is semantic sloppiness, a condition caused by inattention to questions about meaning and inference at the initial stages of inquiry, and which is widespread throughout the history of philosophy. What the logicists learned in developing criticisms of Kant has changed the way philosophy is done: they learned that semantic analysis must be done and done right prior to addressing the questions of philosophy rigorously and systematically.

II.iii Three Strands in Nineteenth-Century Philosophy

This section is a brief overview of three influential philosophical movements in the nineteenth century. The movements are neo-Kantian idealism, positivism, and the semantic tradition; and an instructive dimension along which to compare them is where they stand on *a priori* knowledge.

'Neo-Kantian idealism' refers to a family of schools dedicated to exploring the consequences of Kant's Copernican turn. I use the term broadly, so that it includes other original idealist positions, such as that of Hegel, which are critical of some important points of Kant's philosophy. Perhaps the key tenet that holds these idealist positions together is the view that Kant was right to place synthetic *a priori* judgements at the center of philosophy. Neo-Kantian idealists hold that some truths are knowable *a priori*, and they appeal to the constitutive powers of mind to explain our grasp of them. Within this tradition, the notion that synthesis is the base of all significant philosophy blossoms into magnificent speculative metaphysical systems built upon the synthetic engine of dialectic reasoning.

Positivism is closely related to classical empiricism in creed and in spirit. Positivists shared with their empiricist predecessors a preference for building knowledge from what is given in experience and a distaste for unverifiable speculation. In programmatic terms, the positivists' plan was to shore up empiricism, to steer clear of Berkeley's idealism and to lessen the bleakness of Hume's skepticism, with a rigorous commitment to the scientific method in philosophy. At least implicitly, many empiricists grounded *a priori* knowledge in analyticity. This is exemplified with Hume's category of relations of ideas—the only other sort of knowledge claim, in addition to statements of observable fact, which he considered admissible—on the plausible ground that Hume holds such truths to be knowable *a priori* because they are analytically true.

However, many positivists took Kant's work in this neighborhood to show that, given the possibility that 'analytic' and *a priori* are not co-extensive, analyticity cannot ground the *a priori*. Others were skeptical that such vague concepts had any role to play in a rigorous, scientific philosophy. The official view of positivism is either to say nothing on the subject of *a priori* knowledge or to deny its possibility. Mill (1830) took the latter course, arguing that even the truths of arithmetic are *a posteriori* inductive generalizations. Mill's view on the nature of mathematical truth—which is uncharitably, but perhaps not unfairly, characterized by Frege (1884) as "pebble and gingerbread arithmetic"—is one of the prevalent contenders, along with Kant's synthetic *a priori* view, in opposition to which the logicist thesis is defined.

The proponents of the semantic tradition believe that there are *a priori* truths, but are not at all satisfied with Kant's explanation of our grasp of them. They find positivists' views on many subtle and important questions to be crude and downright incredible, but could not in good conscience appeal to Kant's mysterious faculty of pure intuition to explain where the positivists go wrong. The term 'semantic tradition' is Coffa's (1991); I cite from his elucidation of it:

> The semantic tradition consisted of those who believed in the *a priori* but not in the constitutive powers of mind. They also suspected that the root of all idealist confusion lay in misunderstandings concerning matters of meaning. Semanticists are easily detected: They devote an uncommon amount of attention to concepts, propositions, senses—to the content and structure of what we say, as opposed to the psychic acts in which we say it. ... The basic assumption common to all members of that movement was that epistemology was in a state of disarray due primarily to semantic neglect. ... In particular, they thought the key to the *a priori* lay in an appreciation of the nature and role of concepts, propositions, and senses. (Coffa 1991: 1-2)

Coffa identifies Bolzano and Helmholtz as key early figures in the semantic tradition. Toward the end of different projects, the work of Bolzano and Helmholtz have at least three crucial things in common: (i) both find Kant's views on *a priori* knowledge thoroughly unsatisfactory; (ii) they think that Kant's bedrock examples of synthetic *a priori* judgements in mathematics are based on fundamental misunderstandings; and (iii) they believe that the way to bring these misunderstandings to light, to demonstrate the unsatisfactoriness of the wide-ranging doctrines they support in the Kantian edifice, is to engage in careful semantic analysis. Coffa holds that Frege and Russell are principle proponents of the semantic tradition, and a key reason for this is that the logicist project falls squarely within these three confines.

An essential pillar of the semantic tradition is the view that the key to understanding *a priori* knowledge is a better understanding of meaning. This is not to say that all semanticists are opposed to the very idea of synthetic *a priori* judgements: Frege (1884: 101), for instance, held that the truths of geometry are synthetic *a priori*. Rather, they hold that there is an explanatory connection between meaning and *a prioricity*. In this they agree with Kant's tenet that all analytic truths are knowable *a priori*, that propositions that are true in virtue of meaning can be grasped without checking the facts. However, they dispute Kant's delineation of the set of analytic truths—especially in the case of mathematics—and they seek to go beyond his rather thin conception of analyticity. Semanticists submit the notions of meaning and analyticity to more detailed and rigorous study than did the empiricists or Kantians. To say that some statements are knowable *a priori* because they are true in virtue of meaning is the beginning, not the conclusion, of a philosophical inquiry. It forces questions as to precisely in what such notions as 'truth,' 'meaning,' and 'in virtue of' consist.

Proponents of the semantic tradition hold that semantic analysis is a necessary precursor to any systematic inquiry, that the inferential relations between meanings, concepts, and propositions tops the list of philosophical concerns. Now on to exactly how Frege and Russell came to be the principle proponents of this tradition.

III. The Logicist Thesis

III.i The Foundations of Mathematics

There is a prevalent view of the history of philosophy according to which Frege and Russell mark the end of the modern period, which Descartes had ushered in in the seventeenth century. Descartes' distinctive prioritizing of philosophical concerns had afforded a new and deeply influential paradigm of philosophical inquiry, effectively dethroning the scholastic metaphysical concerns which had dominated the medieval period, and establishing epistemology as first philosophy. Similarly, this view has it, Frege and Russell redraw the confines of the discipline by putting at center stage—by specifying as step one toward any systematic inquiry—questions about meaning and truth and the structure of argument. Throughout much of the twentieth century, in much of Western philosophy, philosophical logic is first philosophy; and that is due to a paradigm shift instituted principally by Frege and Russell, among many others.

There is clearly much to be said in favor of this view. An awful lot of what they were up to, from the point of view of either method or content, did not bear much resemblance to what had come before in philosophy, and has greatly influenced much of what has happened since. Several important works of Frege and Russell come at the beginning of, and are among the most important contributions to, the gradual displacement of epistemological for semantic questions as the primary concern of mainstream academic philosophy. However, one must be careful to avoid overestimating the extent to which Frege and Russell depart from the modern philosophical tradition: understandably enough, much residue of modern philosophy is palpable in the works of both authors, in various ways. That is one of the points that will become clear in this section, via an exploration of their work on the foundations of mathematics.

One fundamental strand of continuity between logicism and the modern tradition is that the questions that logicism is intended to answer are questions of justification, questions about the legitimacy of our claims to knowledge. Logicism is, in part, a thesis about the grounds of mathematical truth, about the grounds of our claims to know things about mathematical objects. The project of showing that mathematics reduces to, or is contained in, logic was undertaken to provide a foundation for mathematics, and there are some significant relations between the logicists' desired foundation and Descartes' foundationalism. My first aim is to explore some of the questions which logicism is intended to answer, and the historical context in which they arose and were perceived to be so pressing. Then in sections III.ii and III.iii I give an overview of the similar projects independently undertaken toward their solution by Frege and by Russell.

It must be noted at the outset that not all explorations into the foundations of mathematics were undertaken in search of answers to exactly the same questions. There are many different issues and questions and strands, which became pressing in the nineteenth century, tangled up under the heading 'foundations of mathematics.' My aim here is to describe, in a general way, two roads into these inquiries. I make no claim that these are the only important aspects of the project—were I writing a different essay, I might well choose to highlight some quite different things. The aspects of the project that I will explore concern questions of form and questions of content. Questions of form concern patterns of inference: for instance, 'How do you know that P follows from, or is supported by, some other proposition Q (or by some other propositions R and S and ...)?' The questions of content, in contrast, concern the meaning of individual premises or conclusions: to wit, 'How do you know that P?' 'What, precisely, does P mean?' As a result of too much assuming and not enough careful justification with respect to these sorts of questions, many—Frege the most vocal among them—felt that the edifice being built up by nineteenth-century mathematicians was resting precariously on shaky ground, ready for an ill wind to blow asunder the house of mathematics.

[1] *Questions of form.* The formal issues center around the question: How do you justify the claim that an inference is formally valid, is good simply in virtue of the structural relations among the parts, and irrespective of their content? The general outline of the formal story can be told with reference to Euclid or to Aristotle: the conventional way to answer such formal questions, until well into the nineteenth century, was to appeal to the authority of one of these thinkers. However, during that century, a number of developments undermined this way of thinking. Here I will explain these developments with reference to Euclid; a similar story about Aristotle is told in section IV.i, in a sketch of the history of logic.

For millennia, Euclidean geometry was taken to be the very paradigm of unassailable demonstrative reasoning, and Euclid's axiomatic method served as the blueprint for how to be certain about a subject matter. A Euclidean thought-process begins from self-evident axioms, takes only obvious and unassailable steps from them, and the self-evidence which characterizes the axioms and rules of inference is inherited by any conclusions drawn. Neither axiom nor steps nor, therefore, conclusions—so the litany goes—could conceivably be false. The first of the blows to befall this line of thought in the nineteenth century came with the development of non-Euclidean geometries. Alternative geometrical systems were developed which denied Euclid's axiom that parallel lines do not meet, the result being a number of consistent and useful, but incompatible, geometrical systems. There was no standard by which to answer the question of which system is true, and this raised unprecedented questions about the notion of mathematical truth. Regardless of whether one finds it self-evident that parallel lines do not meet, it may not be true. Competing claims about parallel lines can each occur to different people as self-evident, but surely they can not all be true. More deeply, what one finds self-evident depends on what one believes, but the truth of the matter in mathematics (according to most mathematicians, Frege the most vocal among them) does not depend on anyone's beliefs.

So, in the wake of the development of non-Euclidean geometries, philosophers and mathematicians had to face for the first time difficult sorts of question about how one might justify the claim that an axiom or rule or statement is true. It is no longer self-evident what is or is not self-evident, or even what that might mean, precisely.

Another blow to this appeal to self-evidence came with the derivation of a number of paradoxes and antinomies, of some rather unsettling results via proofs that were thoroughly unobjectionable by contemporary standards. Cantor's proof that the set of natural numbers has the same cardinality as certain of its proper subsets deserves mention here, as do some of his other provocative results about infinite numbers. Burali-Forti's and Richards' are among the more famous of a number of derivations of contradictions from seemingly innocuous premises. Russell's paradox (which will be described in depth in section V.iii) hit especially hard, as the only assumptions relied on in the derivation of the paradox are what many would classify as analytic truths, constitutive of what logicians and mathematicians mean by the term 'set.' Is the notion of a 'set'—i.e., a collection of entities that share some feature in common—incoherent? Is there some problem with Russell's rather straightforward derivation? Either option would come as quite a blow to mathematics, and no alternative proposal met with wide acceptance.

This first road into the foundations of mathematics concerns not the content of the axioms or premises, *per se*, but rather concerns the way in which the parts of a system support each other, from a formal point of view. With respect to the foundational epistemic question: What grounds your claim to know something? the appeal to self-evidence is undermined, both in the case of justifying a choice of axioms, and of justifying steps in the chain of inference. So, if claims to validity in mathematics cannot be justified in these traditional ways, then how can they be justified? Feelings of self-evidence are no sure guide to truth, let alone to the stronger notions of validity or certainty. Most mathematicians and philosophers think that Kant's appeal to intuition will not do for such solid objective purposes. Neither consistency, coherence, clarity, nor distinctness can provide the requisite ground. So what can?

At the root of this formal road into our foundational questions are questions about the notion of proof, and the logicists hold that developing a more rigorous theory and system of deduction is the best way to respond to the challenge. Important early steps toward a more precise notation in which more varied patterns of inference could be studied were taken by Boole (1854), among others; this line of inquiry blossoms into the deductive systems developed by Frege, Peano, and Russell and Whitehead, which will be described in the next two sections. The following excerpt from Frege gives a sense of the logicists' strategy for responding to this formal challenge:

> The ideal of a strictly scientific method in philosophy, which I have tried to realize here, and which might perhaps be named after Euclid, I should like to describe in the following way. It cannot be required that we should prove everything, because that is impossible; but we can demand that all propositions used without proof be explicitly mentioned as such, so that we can see distinctly what the whole construction rests upon. We should, accordingly, strive to diminish the number of the fundamental laws as much as possible, by proving everything that can be proved. Furthermore I demand—and in this I go beyond Euclid—that all modes of inference must be specified in advance. Otherwise it is impossible to ensure satisfying the first demand. (Frege 1903: 137)

The ultimate positive effect of this challenge was that it led to a crisis about how to justify steps in the chain of inference, and thereby precipitated a renaissance in the study of logic. Mathematicians and logicians began to unearth the assumptions implicit in, and to better understand the nature of, their ancient disciplines.

[2] *Questions of content.* The questions about content that spurred inquiry into the foundations of mathematics were themselves spurred by the unprecedented progress that was made in the study of mathematics in the nineteenth century. To cite some of the highlights: imaginary numbers were shown to be essential for representing motion in a plane and incorporated into a general theory of complex numbers; more or less rigorous definitions were supplied for negative and irrational numbers; the logical basis of the real number system was described; and Cantor's work on the theory of infinite numbers charted new territory to be explored by the mathematical imagination.

Frege alleges that these luxuriant expansions in the house of mathematics are inadequately supported. Many of the developments rely on our grasp of the natural numbers, which are taken as the basis for defining further concepts of arithmetic and analysis. However, according to Frege, we do not sufficiently understand the naturals, and therefore cannot in good conscience reduce other problems to them. In the following passage, he observes that Weierstrass, Heine, and Cantor define the concept of number in quite different ways, and continues:

> Clearly each of them associates a different sense with the word 'number.' The arithmetics of these three mathematicians must therefore be radically different from each other. A sentence for one of them must express a proposition entirely different from the proposition expressed by an identical sentence of the other. (Frege 1979: 215)

On the fundamental question, 'What is a number?' Frege sees that there is no consensus among mathematicians. He complains to Hilbert that "anarchy and subjective inclination reign supreme" (1980: 62) on such deep abstract conceptual matters. This is a lack of foundations, in a different sense: there are no solid and rigorous answers on offer to crucial and fundamental questions concerning what mathematics is about, what mathematics is the science of. A large part of the logicists' project is dedicated toward rectifying this situation. Here again is Frege, writing on the concept of number:

> Yet is it not a scandal that our science should be so unclear about the first and foremost among its objects, and one which is apparently so simple? ... If a concept fundamental to a mighty science gives rise to difficulties, then it is surely an imperative task to investigate it more closely until these difficulties are overcome. ([4]: 136)

So, as it struck Frege, and a little later on Russell, the content-foundational issues concern such questions as: What is a number? What are the objects of mathematical inquiry? Precisely what is mathematics about?

One moral which Frege and Russell draw from their thinking about these questions is that only a realist theory of the content of mathematics will do: the objects of mathematical inquiry must exist mind-independently, and proving truths about them constitutes genuine discoveries, as opposed to building up constructions or laying down conventions. Prevalent alternative candidates to a realist view include psychologism (roughly, the objects of mathematical inquiry are ideas or images in the minds of humans) and formalism (roughly, the objects of mathematical inquiry are numerals and the conventional rules which govern their manipulation). Frege amasses a battery of arguments that these approaches do not stand up to thorough scrutiny, that they are too superficial to provide a solid ground for the truths of mathematics; Russell takes up, extends, and amplifies this battery. Historically (although not conceptually—proponents of psychologism or formalism might still see some virtues in reducing mathematics to logic) logicism goes hand in hand with the realist view that numbers just simply cannot be private psychological property, or mere symbols whose powers and properties are conferred on them by us. Numbers must be real mind-independent objects, in order to explain our intersubjective grasp of arithmetic truths, to explain the efficacy of mathematical laws in keeping bridges and buildings aloft.

In this way logicism leads to the question of precisely which mind-independent objects numbers are, and that question is taken up by Frege's (1884) *Foundations of Arithmetic* and by Russell's (1903) *Principles of Mathematics*, two seminal works in philosophical logic. These works are about the objects of the study of mathematics; but, because of the distinctive way in which those questions are formulated and addressed, they lay much of the groundwork for the philosophy of language.

In this second way into the foundations of mathematics, the questions center not on inference and proof, but rather on meanings and concepts, what they are and how they relate. With unprecedented progress in mathematics, there are widening divergences (at least implicitly) concerning fundamental questions such as: What is a number? What is mathematics about? In the intellectual careers of both Frege and Russell, this state of affairs leads first to complaints about the semantic sloppiness of contemporary mathematicians; then to convictions about a lack of depth and substance in some of the prevalent views about mathematics; and finally to inquiries into the foundations of mathematics, to the project of using the laws of logic to cement the foundations. These questions concerning what mathematics is about are in some sense prior to formal questions about the notion of proof: they are about the objects and properties referred to in the premises, rather than about what this or that set of premises does or does not support. Key successors and contemporaries of Frege and Russell, with respect to these content-foundational questions, include Cantor, Dedekind, and Peano.

To sum up: Cartesian foundationalism, that fountainhead of modern philosophy, is an epistemological project whose aim is the justification of claims to knowledge about the empirical world, the refutation of a certain kind of skeptic. I have isolated and discussed two strands of the logicists' explorations into the foundations of mathematics: questions about the logical foundations of mathematics and foundational questions about the content of mathematics. The former focus on formal patterns of inference, the latter focus on precisely what mathematics is about. These foundational projects diverge from Descartes' not so much in where philosophers want to go but in how to get there. The departure from Descartes' philosophical program instituted by Frege and Russell consists largely in distinctive and insightful ideas about how to achieve their goals. It is not so much that they find Descartes' epistemological questions ill-conceived or unimportant; it is rather that they take questions about meaning and inference to be essential first steps in any systematic philosophical inquiry. In this sense, the logicists' displacing of the Cartesian paradigm has more to do with method than substance. Their place in history is secured by identifying and recognizing the importance of these semantic questions, and in the tools, methods and distinctions they develop to deal with them.

Precisely where these strands converge, where foundational inquiries concerning logic and content meet, is in the project of philosophical logic. It is not unfair to characterize the logicists' answer to these challenges about the ground of our grasp of the truths of mathematics as the claim that these are analytic truths, that they are true in virtue of meaning—(cf., e.g., Russell ([15]: 295-96).) The study of formal deduction is an attempt to get more rigorous about the 'in virtue of' relation, about the precise relation between the laws of logic and the patterns of inference required for mathematics. The study of content is an attempt to get more rigorous about the meanings of some relevant notions. The foundations of mathematics are to be shored up by establishing analytic links in the chain of inference, and analyticity depends on both form and content.

Hence, Frege and Russell hold that much work has to be done in both philosophy of language and the formal study of inference in order to answer some of these questions about our knowledge of mathematics. A conclusive proof of the logicist thesis must include both a better system of deduction (i.e., a rigorous notion of proof) and a realist theory of the content of mathematics (i.e., a rigorous account of precisely what the relevant premises mean).

III.ii Frege's Program

The most productive years of Frege's career, his most innovative work, was dedicated to establishing the logicist thesis. Frege believed that the science of mathematics would not be free from the threat of error or contradiction until a gapless proof of each and every step, based solely on the laws of logic, could be given. Here is a characteristic excerpt, from the early pages of his attempt to define the concept of number:

> The problem becomes, in fact, that of finding the proof of the proposition, and of following it up right back to the primitive truths.... [The] fundamental propositions of arithmetic should be proved, if at all possible, with the utmost rigor; for only if every gap in the chain of deductions is eliminated with the greatest care can we say with certainty upon what primitive truths the proof depends.... (Frege 1884: 4)

His goal was to systematize mathematics, completely and rigorously, by showing that all arithmetical notions could be defined in terms of notions required for the study logic, and that all arithmetical inferences could be derived from the laws of logic.

The aim of the logicist project is to determine "... how far one could get in arithmetic by means of logical deductions alone, supported only by the laws of thought" ([1]: 93). The plan was to develop first an artificial language in which gapless proofs could be constructed (*Conceptual Notation*, 1879). The next step was to define the basic concepts of arithmetic purely logically (*Foundations of Arithmetic*, 1884). The final step was to consist of proofs of the fundamental laws of arithmetic based on the (1884) definitions, within the (1879) system of inference (*Basic Laws of Arithmetic*, volume 1 in 1893, volume 2 in 1903, never completed as planned). However, Russell's paradox permanently derailed the project—see section V.iii for more on this.

In his attempt to establish that the laws of arithmetic rest upon only the laws of logic, before he could begin to construct proofs, Frege encountered some semantic problems. For instance: he saw the need to draw some distinctions more finely than his language would permit, he wanted to express things for which there exists no exact linguistic expression, and so on. These semantic problems can be helpfully split up into two sorts: the first have to do with the computational powers of traditional logic, and the second have to do with the expressive resources of natural language. The computational problems have to do with identifying patterns of inference that are crucial to mathematics but yet go beyond the means of traditional logic. The expressive problems arise because Frege perceived a need for a degree of precision that cannot be attained in what he calls ordinary language—i.e., everyday all-purpose languages such as English or German. In the next few paragraphs I will concentrate on these latter expressive limitations, on what Frege calls the logical imperfections of language. The former, the computational problems Frege encountered within the confines of traditional logic, will be addressed in section IV.i. (It will there become evident that the above computational/expressive distinction is somewhat artificial. These aspects can be split apart in the abstract, but they cannot be so easily separated out in any individual case, because particular computational problems are integrally tied to particular expressive limitations or infelicities.)

Frege describes a thread running through his career as follows:

> I started out from mathematics. The most pressing need, it seemed to me, was to provide this science with a better foundation.... The logical imperfections of language stood in the way of such investigations. I tried to overcome these obstacles. In this way I was led from mathematics to logic. (Frege 1979: 253)

More specifically, the reason he had to re-form logic in order to shore up the foundations of mathematics has to do with certain features inherent in natural languages:

> Language proves to be deficient ... when it comes to protecting thought from error. It does not even meet the first requirement which we must place upon it in this respect; namely, being unambiguous.... Language is not governed by logical laws in such a way that mere adherence to grammar would guarantee the formal correctness of thought processes. The forms in which inference is

expressed are so varied, so loose and vague, that presuppositions can easily slip in unnoticed.... We need a system of symbols from which every ambiguity is banned, which has a strict logical form from which the content cannot escape. ([2]: 120-21)

In short, the sort of rigor demanded for the logicist project goes well beyond the expressive resources of natural languages. A new medium is needed for the study of inference.

Examples of what Frege calls 'the logical imperfections of language' are ambiguity (i.e., some expressions have more than one meaning), vagueness (i.e., some expressions lack any precise meaning), and context-sensitivity (i.e., the meanings of some expressions vary from context to context). These features are widespread in natural languages, which are all-purpose instruments, usable for an uncountable variety of tasks, whose form and content are shaped by all manner of historical accidents and socio- political pressures. There are, however, contexts in which precision is necessary, in which these loose connections between expression and meaning are debilitating barriers. Artificial languages are introduced for these contexts. (One familiar example of an artificial languages is the language of elementary arithmetic, which consists only of the precise operations of addition and multiplication, the identity predicate, and the natural numbers. Another is legal jargon, in which such features as ambiguity, vagueness, and context-sensitivity are avoided by stipulation—e.g., "For the purposes of this document, 'tenant' is to mean ..., 'landlord' is to mean....")

Artificial languages are designed for a special purpose, crafted toward the end of precision in some particular context, and in such a language it is possible to eliminate any feature of natural language that one considers undesirable. Frege insists on the need for an artificial language for the study of logic, in order to overcome some of the semantic and logical obstacles that were encountered in the late nineteenth century. That language is developed in *Conceptual Notation*. (Note that we are getting to the very center of the onion here: the reason why mathematics, logic, and the philosophy of language were so tightly interwoven around the turn of the twentieth century is precisely that (a) these philosophers of mathematics encounter computational problems in logic and expressive problems with language on the path to their desired ends, and (b) they hold that philosophical logic can set things right.)

Frege employs an analogy between his artificial language and a microscope ([1]: 94). Like the microscope, the conceptual notation is an artificial device created purely for scientific purposes. Like the relation between the microscope and eye, this script is not intended to render natural languages obsolete. It is rather intended to improve our understanding of a very specific range of issues. If you want to look at very very small things, a microscope is to be preferred; otherwise, in day to day affairs, an unfettered eye is incomparably preferable. Similarly, if you want to be rigorous about formal validity, use the conceptual notation; but to order a sandwich or to discuss current affairs, natural languages are much more appropriate.

So the first task in Frege's project was to reform logic, and toward that end he designed an artificial language, a medium of expression created specifically for the study of inference. This language was designed to be the perfect medium for the study of inference, in which every symbol stands unambiguously for one precise concept. It is designed to be a language in which all and only that which is relevant to inferential properties is reflected in the syntax, in which "mere adherence to grammar can guarantee formal correctness of thought processes" ([2]: 120). Thoughts expressed in this notation have a degree of precision that natural language expressions can only approximate, and are better suited for unassailable formal reasoning. The closer we can come to such a transparent link between a sentence and the thought expressed, according to Frege, the more precise will be our reasoning.

One reason why *Conceptual Notation* is such a landmark is that it marks the first application of mathematical notions to semantic problems. Frege radically altered the character of logic, replacing the divide between subject and predicate as its fundamental distinction with that between function and argument. (These innovations will be explained in depth in sections IV.i and IV.ii. A function is a relation that depends upon and varies with its constituent terms, and an argument is an element in the domain of a given function upon which the value of the function depends.) No less than eleven distinctive elements of modern logic, from quantifiers and variables to the propositional calculus as a logistic system, were first formulated in that work.

Frege's *Conceptual Notation* is perhaps the single most important contribution to a tradition known as the ideal language project. Leibniz is often accredited as the originator of this tradition, but a number of loosely related projects in the neighborhood constituted something of a popular Enlightenment hobby. There were several proposals to create a language that is in some respect universal or perfect. Undeniably, though, in a series of papers written from 1679-1690, Leibniz did some characteristically deep and interesting work on this notion of an ideal language. He described his goal as an alphabet of human thought, a universal language or script in which all truths of reason could be reduced to a calculus, a system of expression in which every symbol transparently and unambiguously stands for one concept. Russell quotes Leibniz approvingly:

> If controversies were to arise, there would be no more need of disputation between two philosophers than between two accountants. For it would suffice to take their pens in their hands, to sit down at their desks, and to say to each other (with a friend to witness, if they liked), 'Let us calculate.' ([10]: 224)

This notion of an ideal language is worth a look, as Frege and Russell (as well as many of their successors) certainly see themselves as contributing to this ideal language tradition.

There are two aspects to Leibniz' project: the *calculus ratiocinator* (or calculus of thought) and *lingua characteristica* (or universal language). A *lingua characteristica* is a language in which everything conceivable can be clearly and unambiguously expressed; a *calculus ratiocinator* is a language designed for making inferences, doing proofs, calculating relations among thoughts. The *lingua characteristica* is an expressively ideal vocabulary, while the *calculus ratiocinator* is computationally ideal grammar. Leibniz' goal is to first develop the lexicon, the alphabet of thought, and then to develop the purely rational grammar, in which adherence to grammar will guarantee correctness of thought process. The result would be an infallible instrument of deductive proof. (It is hard to find a more extreme example of Enlightenment optimism about the prospects of human reasoning!) This is a goal which Frege reveres, but he holds that Leibniz' goal is too ambitious to be attained in one step. However, Frege holds that progress has been made toward this end, and that there is reason to believe that more progress will follow. Frege's statement of how his work fits into this tradition is worth quoting for, among other things, its illustration of the kind of patient and careful analytical methodology that would come to characterize his enduring contribution to philosophy:

> But even if [Leibniz'] high aim cannot be attained in one try, we still need not give up hope for a slow, stepwise approximation. If a problem in its complete generality appears unsolvable, we have to limit it provisionally; then, perhaps, it will be mastered with a gradual advance. We can view the symbols of arithmetic, geometry, and chemistry as realizations of the Leibnizian ideal in particular areas. The [conceptual notation] offered here adds a new domain to these; indeed, the one situated in the middle adjoining all the others. Thus, from this starting point, with the greatest expectation of success, we can begin to fill in the gaps in the existing formula languages, connect their hitherto separate domains to the province of a single formula language and extend it to the fields which up to now have lacked such a language. I am sure that my [conceptual notation] can be successfully applied wherever a special value must be placed upon the validity of proofs.... ([1]: 95)

Frege sees himself as taking a big step toward Leibniz' seminal goal. However, *Conceptual Notation* did not get rave reviews—it barely got reviews at all. Most of those who noticed the work dismissed it as a bad version of what Boole had done more clearly. (See [3] for Frege's address to this criticism—at best, Boole had attempted a *calculus ratiocinator*. Frege shows the limits of Boole's work toward that end; and further, he had attempted things Boole could not have conceived of, on the *lingua characteristica* arm.) Fundamentally, the reason why that monumental work went unappreciated for so long is that nobody had understood it. The book was full of bold innovations and strange two-dimensional proofs in an alien script. Even Russell, who was not only uncommonly intelligent but was in complete agreement with Frege on several important relevant points, said that "owing to the great difficulty of [Frege's] system, I had failed to understand its importance or to grasp its contents" (1903: xvi). Part of the reason why Frege spent his entire career in obscurity lies in his not having explained, in accessible terms, his strange, cumbrous two-dimensional notation. Such a clarificatory introduction is all the more important in this case, because the motivations for the project were out of step with contemporary trends, and because the content of the work was so unlike anything ever seen in philosophy or logic or mathematics. There just simply was no audience that understood what Frege is up to, in [1].

III.iii Russell's Rediscovery of Logicism

Russell is one of the most widely read and influential philosophers of the past few centuries. He is a prolific writer, whose works range over a thoroughly diverse terrain—he left unanalyzed few aspects of human thought, experience, and culture. However, Russell's reputation was originally staked on work in the philosophy of mathematics. His deepest and most significant contributions to philosophy, which came in the early decades of the twentieth century, were spurred by logical, semantic, and epistemological questions that grew out of his and others' research in the philosophy of mathematics.

Russell reports in his *Autobiography* that his interest in the foundations of mathematics was kindled at the age of eleven, when his older brother introduced him to the works of Euclid, and young Bertrand wanted to know what ground or justification there is in support of the axioms. It seems that no one could give him a satisfactory answer, and so he dedicated much of his mental energy over several decades to providing one. The philosophy of mathematics he would arrive at is, at many points, indistinguishable from Frege's. The logicist thesis lies at the heart of many of his works, including *Principles of Mathematics* (1903)—"The fundamental thesis of the following pages [is] that mathematics and logic are identical" (1903: p.v)—and his and Whitehead's *Principia Mathematica* (1910), whose goal is "the deduction of pure mathematics from its logical foundations." (1910: 90) In the preface to that work he and Whitehead acknowledged a debt to Frege "in all questions of logical analysis." Further, Russell agrees with Frege not only about the relation between logic and mathematics, but also about the issues which pertain to content, about the questions which concern what mathematics is about: he says, "The philosophy of arithmetic was wrongly conceived by every writer before Frege." (1959: 68) Russell, too, pursues philosophical logic as a means to his ends: "... the study of grammar ... is capable of throwing far more light on philosophical questions than is commonly supposed...." (1903: 42) Quite akin to Frege's project, Russell's project embraced both formal considerations and aspects that pertain to content, both the definitional work of philosophical logic and the derivational work of formal logic. In the course of his efforts to demonstrate the logicist thesis, Russell is quite on par with Frege in establishing the philosophy of language as an autonomous subject-matter.

Among the most striking things about the similarity among the views of Frege and Russell is the fact that they arrived at these views independently. Russell reports in the preface to his (1903) *Principles of Mathematics* that he had been aware of Frege's work for some time, but he did not understand it until after he had arrived at a similar position. That work was already in press before Russell grasped the import of Frege's writings. Russell included an appendix on Frege's thought, acknowledging Frege's "discovery of propositions which, when [Russell] wrote, [he] believed to have been new." (1903: 518)

It was Peano, whom Russell met in 1900 at an International Congress of Philosophy, who set Russell on the path to logicism. Influenced by Boole's development of a symbolic algebra of logic, Peano had invented a notational system consisting of a small range of symbols that stood for what he took to be the essential concepts of arithmetic. He had been working at rewriting different fragments of mathematics in his symbolic notation—the most enduring product of his efforts being the Peano Axioms—and the hope was to arrive at a system in which all of mathematics could be expressed. Russell's meeting Peano caused him to undergo what he himself characterized as a revolution. He was overwhelmed by Peano's "precision" and "logical rigor," (1959: 65) and was struck by the promise of Peano's notation to make perspicuous the philosophical foundations of mathematics.

The revolution came with the thought that the basic concepts for which Peano's symbols stood could be reduced to purely logical notions, and thus that all the assumptions and laws needed for mathematics were provided by logic. If Peano is right that we can formalize or express all of mathematics in a logical notation, then perhaps the laws of logic would suffice to justify all of the inferences in mathematics. Hence Russell's rediscovery of the logicist thesis: the foundations of mathematics, over which he had lost sleep since adolescence, could be guaranteed by the laws of logic. To illustrate: one of Russell's basic insights (which echoes Frege's (1884) strategy almost exactly) is to add the notion of a set to Peano's axioms. This would permit a definition of 'number,' in terms of equinumerous sets; and then this definition of number could form the basis for definitions of other requisite notions of arithmetic. From there, a proof of the analyticity of arithmetic, a refutation of Kant's and Mill's views of mathematics, would show that the laws of logic are the only modes of inference required. This is a dream that Frege shared, and that is paradigmatic of much work in the semantic tradition: a conclusive disproof of Kant's view about the synthetic nature of mathematical truth, via semantics, and that via careful analysis of mathematical examples.

Peano was one of the few people who had reviewed Frege's *Conceptual Notation*. His unfavorable review was almost certainly not disinterested. Peano claimed that his system was clearer and more precise than Frege's, and that it was better suited for the formalization of mathematical proofs. Although it seems evident that

Peano had not really understood Frege's system, his was an important review, because it initiated a fruitful correspondence between the two. Frege eventually convinced Peano that his own notation, although more forbidding and cumbrous, is much better suited for the derivation of mathematics from logic. It was Peano who directed Russell back to Frege's works, and Russell concludes from the comparison that:

> [Frege's] symbolism, though unfortunately so cumbrous as to be very difficult to employ in practice, is based upon an analysis of logical notions much more profound than Peano's, and is philosophically superior to its more convenient rival. (Russell 1903: 501)

Russell and Whitehead (1910) took on the basic notions of Frege's logic, simplified the symbolism somewhat, and set out to derive conclusively mathematics from logic. One of the many positive effects of that monumental work is that modern logic is therein systematized and standardized.

Another factor that is crucial to Russell's rediscovery of logicism was the influence of G.E. Moore, and their celebrated revolt against idealism. Moore was a student with Russell at Cambridge, and the prevailing atmosphere there during their studies, in the 1890s, was predominantly Hegelian. Both Russell and Moore were under the spell of idealism for a time. Key catalysts in their rejection of idealism were the writings of a widely influential central-European realist movement, a faction within the semantic tradition. (The most important figure here, for the purposes of this story, is Meinong, whose views will be discussed in section V.i.) Russell's and Moore's conviction that the root of many idealist confusions lay in misunderstanding matters of meaning echoes sentiments raised and explored by their central-European realist predecessors. This influence resonates in Russell's claim that "faulty logic and excessive reliance on the subject-predicate form are the basis of idealism" (Coffa, 1991: 95), and throughout Moore's (1903) "Refutation of Idealism." Russell's "distinction between the act and the object in our apprehending of things" (1912: 42)—which is crucial not only to the rejection of idealism, but also to the creation of the philosophy of language as an autonomous area of inquiry—is perhaps the defining tenet of this realist movement. Further, like Meinong, Russell and Moore both endorsed rather extreme realist positions for a short time around the turn of the century.

Rejecting idealism propels Russell toward another pillar of Frege's logicism: a realist theory of mathematics. Russell could never quite stomach the degree of psychologism about mathematics that is inherent in idealism. In this respect, he refers to his rejection of idealism as a liberation from a subjective prison: "Above all, I no longer had to think of mathematics as not quite true." (1959: 61) Rejecting idealism opens up a mind-independent world out there, to be explored and charted. It brings him hand in hand with Frege, against Kant, on the question of the mind-independent status of the objects and laws of mathematics, as well as on the question of the analyticity of the truths of mathematics.

Among the noteworthy respects in which Russell's logicist project differs from Frege's is that a strong strain of Cartesian foundationalism is much more evident in Russell's thought. Frege would agree with Russell's claim that: "The characteristic excellence of mathematics is to be found where the reasoning is rigidly logical: the rules of logic are to mathematics what those of structure are to architecture." (1902: 63) However, Russell goes beyond Frege in seeing in the foundations of mathematics a way to complete Descartes' ambitious epistemological project: "I hoped sooner or later to arrive at a perfected mathematics which should leave no room for doubts, and bit by bit to extend the sphere of certainty from mathematics to other sciences." (1959: 36) So, while both Frege and Russell pursue philosophical logic to cement the foundations of mathematics, at least initially, Russell has some more general goals. This may go some way toward explaining their different reactions to Russell's paradox, but that story must wait until section V.iii. Let us now turn to the tools and methods of the project of philosophical logic.

IV. Philosophical Logic

IV.i Modern Logic

Logic is the study of formal patterns of inference, of what premises support or justify in virtue of the relations among their parts, and irrespective of their particular meaning or content. Those patterns of inference that are structurally sound, or good in virtue of the formal relations among their parts, logicians call 'valid.'

Aristotle was the first to study logic, and he based his study on the syllogism. The following are examples of syllogisms:

(1) Humans are mammals, and mammals are mortal. Therefore, humans are mortal.
(2) No dictatorships are democratic, and some corporations are democratic. Therefore, some corporations are not dictatorships.

Both syllogisms are valid, underlain by the valid logical forms:

(1*) All A is B, all B is C. Therefore, all A is C.
(2*) No A is B, some C is B. Therefore, some C is not A.

Aristotelian logic—also called 'syllogistic,' 'categorical,' and 'traditional' logic—is exclusively concerned with cataloguing inclusion and exclusion relations among categories, as these are expressed by syllogisms. Aristotle's is a marvelously elegant and comprehensive system of logic: thinking about any subject can be made more perspicuous and rigorous by applying this kind of syllogistic reasoning. Facility with these syllogistic relations is enormously helpful in judging the validity of arguments.

Aristotle's system is so comprehensive that many felt that it catalogues deductive inference completely, that it would not be possible to improve upon it. As noted in section III.i, this opinion (along with the association of Euclid with unassailable demonstrative certainty) took some serious blows in the nineteenth century. In response to some of the challenges faced by mathematicians and logicians about justifying claims to mathematical knowledge, theorists began to unearth principles of inference central to mathematical proofs that go well beyond the resources of syllogistic reasoning. Some computational limits inherent in traditional logic, and some closely related expressive limitations of natural language, became evident. These limitations propelled Frege and Russell, among others, to develop artificial languages for the study of logic; and this proved to be an enormously fruitful move for the development of logic, and for the study of language.

The central point of the present section concerns these computational limits of categorical logic; and, because of the importance of this point for the developments that are the focus of this volume, the point will be unpacked carefully, and from a couple of different directions. First, I will describe some of the kinds of statements whose content cannot be captured via categorical analysis, but which are indispensable to mathematical reasoning. Second, I will move from there to some more general points about logical form. Third, I will run through a set of examples that further illustrates some of the more important morals here, about how these computational limits of traditional logic relate to expressive limits of natural languages. I develop these points in such detail here because in many ways they constitute the crux of this chapter in the history of ideas. We have arrived at the heart of the matter. More so than any other, this section explains how it is that problems in the foundations of mathematics eventually led to the creation of the philosophy of language as an autonomous discipline.

First to some statements whose content cannot be adequately captured by categorical analysis. To apply Aristotelian logic to an inference, all premises must first be forced into the categorical mold 'S is P.' However, several sorts of statements crucial for proofs in mathematics simply do not admit of such analysis. (Not surprisingly, Leibniz recognized this point. He did work to rectify the situation, but his solutions were not as radical as those of Frege and Russell—they were attempts to patch up the old system, rather than to create a new one—and unfortunately much of Leibniz' logical work remained unpublished until quite recently, and so did not have its due influence on succeeding logicians.) One type of problem here concerns utterances containing more than one quantified noun phrase. (A quantified noun phrase consists of a common noun plus a determiner, such as 'no sheep,' 'all Texans,' or 'some but not all Indian restaurants.') Consider, for example:

(1) Every number is greater than some number.

(2) No number is greater than every number.

The best categorical candidates for the logical form of these sentences are:

(1*) All S is P

(2*) No S is P

However, this just simply fails to capture some important aspects of (1) and (2). In (1) and (2), there is a tight formal or structural connection between the grammatical subject (i.e., 'number') and predicate (i.e., 'number greater than some/every number') and this is just not reflected in (1*) or (2*). The only two relations between categories that can be formalized within categorical analysis are identity and distinctness; and so the resources are simply not there to capture this kind of intermediate formal link between terms. So the inferential properties of these utterances including two structurally related quantified noun phrases, the exact roles that they play in linking premises in a chain of reasoning, cannot be captured by categorical analysis.

Indeed, one of the problems highlighted by (1) and (2) runs deeper. It concerns multiple generality *per se*—i.e., any statements containing more than one quantified noun phrase, over and above this problem with formally connected quantified noun phrases. Consider:

(3) Every boy loves some girl.

Note first that (3) is ambiguous between a generalization about every boy, that they all love some girl or other, and an utterance about some girl—Mary, for instance—to the effect that every boy loves her. (Any sentence with more than one quantifer is subject to a similar ambiguity, although they are not all equally natural readings. It takes a bit of forcing to get a natural-sounding second reading out of (1), and even more so for (2).) Categorical logic can come up with two readings of (3):

(3*) All S is P (i.e., Every boy is such that ...)

(3**) Some S is P (i.e., There is a girl such that ...)

However, as is explained more thoroughly in section IV.ii, the ambiguity is much better accounted for via the understanding of quantified noun phrases that Frege and Russell prescribe. Categorical analysis can only handle one quantifier per utterance; whereas there is no limit to the number of quantifiers within the resources of modern logic. The following are examples of the phenomenon instanced in (1)-(3), concerning multiple generality, which can be analyzed within modern logic, but not within categorical:

(3a) Every boy is always in love with some girl.

(3b) Everywhere you go, every boy is always in love with some girl.

Modern logic provides a better means of breaking into an utterance in order to formalize the relations among its parts. In due course, we will come back to (3a) and (3b) to prove this point.

Other closely related problems concern statements involving certain kinds of operator expression, such as:

(4) If a number is even, then the square of that number is even too.

(The operator here is 'the square of...'. The term 'operator' will be defined in section IV.ii.) Such expressions are common in natural language, and indispensable to many mathematical proofs, but the intimate relation between the categories 'a number' and 'the square of a number' goes beyond the resources of categorical analysis. Categorically speaking, (4), too, must be cast along the lines of 'All S is P,' which is to miss entirely the formal relation between the expressions 'a number' and 'the square of a number.' Hence, again, categorical analysis clearly does a shoddy job of capturing the content of (4). A new way of capturing content is needed, if we are to study the inferential properties of such sentences.

Most fundamentally, Frege alleges that categorical analysis is confined to an inadequate formalization of a statement as simple as:

(5) 11 has a successor.

One problem concerns the awkwardness of singular terms in categorical logic— the subject expression '11' must be cast as the category 'all numbers identical to 11,' and thus (5), too, is forced into the mold of 'All S is P.' However, Frege's deepest criticism of categorical analysis concerns its inability to adequately identify and represent expressions of generality, not singularity. According to Frege, (5) asserts the non-emptiness of a certain set—i.e., the successors of 11—and, as he puts it at (1884: 100), in such statements, there can be no question of a subject in the traditional sense. 'A successor' is a quantified noun phrase; statements in which such phrases occur are general—are about relations between concepts. The logical form of (5), according to Frege, is:

(5*) At least one thing falls under the concept 'F.'

Here 'F' stands for '( ) is a successor of 11.' Insofar as our aim is to provide a picture of how (5) is apt to link with other premises in a chain of reasoning, this analysis is demonstrably preferable to the cumbersome categorical 'All numbers identical to 11 are numbers that have a successor.' Frege's analysis is not only less awkward; it is, further, an integral part of a much more productive system of logic.

As a final illustration of these computational limits of categorical analysis, consider an instance of existential generalization, a central rule of inference in modern logic:

(6) 11 is odd; therefore, something is odd.

This inference can, with considerable awkwardness, be forced to fit into categorical logic, provided that we add an enthymematic premise:

(6*) All things identical to 11 are things that are odd, [enthymeme] all things identical to 11 are things that exist. Therefore, some things that are odd are things that exist.

On the analysis Frege prescribes, the logical form of existential generalization, which is the inference underlying (6), is:

(6**) x is F; therefore at least one thing is F.

Clearly, Frege's analysis results in a tighter representation of the content of (6). Again, and more importantly, it also constitutes a more fruitful picture of its inferential properties, as is attested by the rapid subsequent development of modern logic.

So, the upshot of this is that traditional logic is ill-suited for the formalization of a wide variety of propositions, and of a wide variety of patterns of inference. Thus arises the need for the study of diverse varieties of logical form, in order even to formulate precisely, let alone to justify, the logicists' thesis about the foundations of mathematics.

Second, let us move to some more general lessons to be learned about logical form and natural language. There can be quite a difference between how a sentence is categorized via the traditional 'S is P' approach and how the proposition the sentence expresses ought to be categorized for the purposes of logical analysis. This gap is manifest in both directions: propositions of the same logical form can be expressed by sentences whose grammatical form is quite different, and sentences of the same grammatical form can express propositions with very different logical forms. The former point is elementary, as there are all manner of familiar cases of formally distinct utterances that say exactly the same thing. For instance, there are active/passive pairs:

(1) Bill missed Mary.
(1a) Mary was missed by Bill.

One can express the same information by interchanging pronouns for certain terms, as appropriate in a given context:

(2) Arthur is sitting.
(2a) I am sitting. [said by Arthur]

Indeed, anyone with any experience with a second language has some degree of facility with the distinction between a proposition and its linguistic clothing. For instance, the following two sentences say exactly the same thing:

(3) I broke my leg skiing.
(3a) Je me suis cassé la jambe en skiant.

So this first point is that formally distinct utterances can correspond to one given logical form, and it gets one halfway to the point of departure for philosophical logic. It establishes this key distinction between the grammatical form of a sentence and the logical form of the proposition expressed.

The correlative point—that sentences with the same grammatical form can express propositions with distinct logical forms—was less evident. This point is more subtle, and deep. Consider the following:

(4) John came down the road
(5) Nothing came down the road.
(6) All the soldiers came down the road.
(7) A soldier came down the road.
(8) The last free soldier came down the road.

Syllogistic logic prescribes very similar analyses for these sentences: the subject-term names some category, and, in each case, the very same predicate makes the very same attribution. They are all taken to be of the form: 'S is P,' the only dimension along which they differ concerns what 'S' stands for in each case. Frege and Russell recognize, however, that there are some radical differences among the inferential properties exhibited by the propositions expressed by the sentences on this list. This is precisely the point of departure for modern logic, for philosophical logic, for the style of analysis that prompts Wittgenstein's: "Distrust of grammar is a first requisite for philosophizing." (1961: 106)

Consider first that both (7) and 'A soldier did not come down the road' can be true. For no other member of this list can one change 'came' to 'did not come' and yield a pair of sentences which can both express truths, without ambiguity and in the same context. So this is an inferential property that is unique to (7), and it shows that 'a soldier' does not function as a name—that the truth-conditions of statements containing quantified noun phrases of the form 'an F' are not integrally linked to anyone in particular. (7) is about soldierhood, not about any specific individual. (A similar point applies to (6)—only in the case of (6) can one change 'came' to 'did not come' and yield a pair of utterances that are both false, without ambiguity and in the same context. This echoes an example employed by Frege ([6]: 168), in establishing the point that expressions of the form 'all Fs' do not name anything.)

Indeed, (4) and perhaps (8) are the only ones whose truth-conditions are tied to any particular entity. (The reason for the qualification 'perhaps (8)' is that the logical form of (8) is a matter of some contention. Crudely put, Frege holds that the logical form of (8) is akin to that of (4), whereas Russell argues that the logical form of (8) is much closer to that of (6). This issue will be explored in section V.ii.) To highlight this difference, compare the inferences:

(4a) John came down the road, John is now in the backyard. Therefore, something that came down the road is now in the backyard.

(5a) Nothing came down the road, nothing is now in the backyard. Therefore, something that came down the road is now in the backyard.

(6a) A soldier came down the road, a soldier is now in the backyard. Therefore, something that came down the road is now in the backyard.

Only (4a) is valid. The invalidity of (5a) is blatant, as those premises by no means entail that there is any one thing to verify the conclusion. (6a) may be less blatant but is no less invalid. What distinguishes it from (4a) is that the two separate premises may be true of two separate soldiers—again, without ambiguity and in the same context—and hence the truth of both premises is no guarantee of the truth of the conclusion. So here we have an inferential property of (4) that (5) and (6) clearly do not share. The logical forms of these three inferences are:

(4*) x is F, x is G. Therefore, something is F and G.

(5*) Nothing is F, nothing is G. Therefore, something is F and G.

(6*) An F is G, an F is H. Therefore, something is G and H.

(4*) is valid, (5*) and (6*) are not. Hence, this is further reason to discriminate among the logical forms of the propositions expressed by these grammatically similar sentences.

There is an inferential property that distinguishes sentences of the form of (7) from those of the form of (4) and (6). It concerns the relation between truth-conditions and existence, and the example must be changed a little to make the point. Consider:

(4b) John will be prosecuted.

(6b) A trespasser will be prosecuted.
(7b) All trespassers will be prosecuted.

In order for (4b) and (6b) to express truths, somebody has to intend to prosecute someone, whereas (7b) can express a truth in a context where no such intention is ever formed with respect to any specific trespasser. (7b) expresses a relation among concepts whose truth does not depend on anything in fact actually falling under those concepts. (In the jargon, singular (e.g., 4b) and existential (e.g., 6b) propositions do have *existential import*, but universal (e.g., 7b) propositions do not.) We can make significant assertions about concepts—about frictionless planes or unicorns or perpetual motion machines, say—and the fact that nothing actual falls under those concepts does not undermine the meaningfulness of these assertions. So here we have another difference among the inferential properties of the grammatically similar sentences (4) through (8)—i.e., (4b) and (6b) do, but (7b) does not, formally entail that at least one thing will be prosecuted.

To sum up, the logical forms of (4)-(7)—leaving (8) for section V.ii—are:
(4**) x is F.
(5**) Nothing is both F and G.
(6**) All Fs are G.
(7**) At least one thing is both F and G.

The differences in their inferential properties are reflected in these differences in logical form, and these different logical forms establish the seminal point that sentences of the same grammatical form can express very different propositions. This highlights the need for an alternative way of studying logical form that is not so closely tied to the grammatical structure of natural languages. Before turning to some more general points about what is at the root of these differences in inferential properties, let me quickly elucidate a third set of examples.

These examples further illustrate these same central points, and are of interest for various historical and conceptual reasons. Consider the following:
(9) God exists.
(10) God is good.
(11) Humans exist.
(12) Humans think.
(13) Humans are numerous.

The thought that (9) and (10) are statements of the same form is integral to the ontological argument for the existence of God. However, from a logical point of view, (9) is dissimilar from (10) in that (9) does not attribute a property to a subject. Existence is not some property, on a par with others like redness or heaviness, which objects can have or lack. (9) asserts the non-emptiness of a certain set—i.e., at least one thing falls under the concept '( ) is God'—and it is a presupposition for attributing any properties to anything in that set. Note also that (10) formally entails (9) and not vice versa. This shows that we have statements with different logical forms here. (10) is a straightforward attribution of a predicate to an individual; whereas (9) asserts that at least one thing satisfies a certain predicate.

Similarly for (11) through (13)—although they look quite similar from a grammatical point of view, they are logically quite distinct. (11) asserts the non-emptiness of a certain set: at least one thing is human. (12) asserts a conditional relation among predicates: that to anything to which the predicate '... is human' applies the predicate '... thinks' also applies. (13) asserts that there are many instances of the concept human: it is an assertion about the number or amount of things which fall under that particular concept. So, the logical forms of these sentences are:

(9*) There is at least one F.
(10*) x is F.
(11*) There is at least one F.
(12*) All Fs are G.
(13*) Many things fall under the concept F.

In working through these examples, particularly the second set, we have arrived at one of the three seminal distinctions that lie at the base of the philosophy of language as an autonomous area of inquiry: the distinction between referring expressions and quantified noun phrases. (The other two are Frege's distinction between sense and reference and Russell's distinction between referring and denoting; these are explored in sections V.i and V.ii, respectively.) I will explain this distinction with reference to sentences (4)-(7):

(4) John came down the road.
(5) Nothing came down the road.
(6) All the soldiers came down the road.
(7) A soldier came down the road.

(5)-(7) express relations among concepts, as opposed to (4), which specifies, and attributes a predicate to, an individual. The point is most obvious in the case of (5): it is a huge mistake to look for a subject—for some thing that the subject-term stands for and upon which the predicate hangs a quality—in this case. But the point applies to all quantified noun phrases, and the subject-terms in each of (5)-(7) are quantified noun phrases. Sentences whose subject-term is a quantified noun phrase express assertions about concepts or predicates; sentences whose subject-term is a referring expression, such as (4), express assertions about individuals. It is important to attend to this distinction in logic and in philosophy more generally, because these two types of utterance have very different meanings, and very different inferential properties.

Compare (4) and (7). What (7) says is true if and only if there is at least one thing that satisfies both of the predicates '... is a soldier' and '... came down the road.' (4), in contrast, is true if and only if John has come down the road. The truth-condition of (4) is entirely a matter of how things are with John, whereas the truth-condition of (7) is not so intimately tied to anyone in particular. (7) is indifferent to which individual makes it true, in a way that (4) is not. Let us use the terms 'object-dependent' and 'object-independent' to mark this contrast. If the truth-conditions of a proposition depend on the states and doings of some one individual in particular, then the proposition is object-dependent. If the truth-conditions of a proposition do not depend on any one in particular, then it is object-independent. (4), and indeed any sentence with a name like 'John' in the subject-position, expresses an object-dependent proposition: such propositions are true or false according to how things are with John. (5)-(7) express object-independent propositions. Even if John is the one and only soldier who makes (7) true in a certain context, the proposition is not so tightly tied to John. Witness the fact that it would still be true if John had stayed home and some other soldier came down the road instead. More importantly, not merely would it still be true; (7) would express exactly the same proposition in such a context. (7) is about soldierhood, not about any particular soldier, regardless of whether there is, in fact, exactly one soldier salient in the context of utterance.

Utterances whose subject-term is a referring expression express object-dependent propositions, and utterances whose subject-term is a quantified noun phrase express object-independent propositions. This object-dependent/object-independent distinction was not understood, and hence not carefully attended to, in traditional logic. Frege and Russell have much to say about the perils that result from such inattention to fine semantic distinctions. Many important subsequent debates within the philosophy of language, including those discussed in sections V.i and V.ii, concern precisely where and how this important distinction should be drawn.

To sum up: On the traditional approach, there are two essential logical constituents of a proposition, and they correspond to the two essential grammatical constituents of a sentence. The subject term names what the sentence is about, and the predicate term hangs a quality on it. Frege and Russell recognize that this traditional approach misses some important distinctions—perhaps most importantly, the distinction between referring expressions and quantified noun phrases. They reject this reliance on the subject/predicate distinction as the fundamental semantic divide, and they proceed to develop a better notation for the study of logical form. Although by no means without insightful and significant immediate predecessors in the study of logic—perhaps Boole and Peano chief among them—it is this move that characterizes modern logic. It is this recognition of the expressive and computational limitations of traditional logic that forces the need for the development of an artificial language for the study of logic, and that precipitates the seminal work in philosophical logic of identifying and cataloguing the diverse varieties of logical form.

Our next question concerns precisely what is to replace the subject/predicate distinction as the fundamental semantic divide. We have already met with the answer: Frege's function-argument analysis of language. Now let us explore where this analysis came from and what its substance is.

IV.ii Function and Argument

The matter we have arrived at may well be Frege's principle innovation. At the very least, it is integrally tied to most of Frege's innovations. The subject-predicate distinction is diagnosed to be too superficial, from the points of view of logic and semantics. In contrast, the function-argument analysis has many virtues. It affords a more fruitful means of isolating and representing what is relevant for the study of inference, and for the study of meaning. It allows a univocal treatment of a number of seemingly disparate semantic phenomena. It is this innovation that allows a language to be treated as a calculus, and that has reaped many benefits. The function-argument analysis of language marks the first application of mathematical tools to the analysis of language, and that is a huge milestone, seminal for any discipline—from cognitive psychology to theoretical linguistics to computer science—that studies formal aspects of information processing and human cognition.

Frege did graduate work in mathematics on the concept of a function, and this is a key factor precipitating his semantic innovations. A function is a certain sort of relation, that takes elements of a domain as arguments and outputs a certain value for each argument. Elementary arithmetic affords some familiar examples of functions, such as addition (i.e., ... + ...) and multiplication (i.e., ... x ...). If you feed in 2 and 3 as arguments, the first will output 5 as the value of that function for those arguments, the second will output 6. The ellipses here mark the fact that functions essentially include gaps, or spaces for arguments. In order for a function to yield a definite value, one must plug appropriate argument-terms—in the above cases, two numbers—into the appropriate spaces. Only when completed by arguments does the function yield a certain definite value.

Functions can have any number of argument-places—'$(...)^2$' is a function of one argument, '... + ...' is a function of two arguments, 'the triangle described by points ..., ..., and ...' is a function of three arguments, and so on up. In general, a function of n-argument places takes n terms as argument and outputs a certain value for those arguments. Typically, the gaps in mathematical functions are marked by variables—e.g., '$x^2 + 3x$' denotes a function from a number to the sum of its square and its triple. Frege [8] argues that these gaps in functions are better represented by brackets than variables—i.e., $(\ )^2 + 3(\ )$—because the use of variables encourages bad semantic pictures. It encourages one to think that a variable is the name of a variable number; but there is no such thing as a variable number: all numbers are some number in particular. As he explains, this is another instance of how a cavalier attitude toward semantics can engender philosophical mistakes, and it is related to the subject-predicate approach to semantics, in that that approach encourages one to find a discrete entity to serve as the meaning of each significant linguistic expression. Subject-predicate semantics encourages one to think, mistakenly, that since a variable can serve as a subject-term, then it must name some thing.

Frege uses the term 'unsaturated' to refer to this essential gappiness of functions. Functions, such as '( )^2' or '( ) + ( ),' are unsaturated in that they do not stand for any thing in isolation, but rather stand in need of completion or saturation by argument terms in order to designate any specific value. In contrast, argument expressions, such as '2' or '(2 + 3)^2,' are saturated: they do stand for specific individuals on their own. Unsaturated function-terms are meaningful expressions that do not stand for any thing in isolation, according to Frege, whereas saturated argument-terms are expressions that do independently stand for some individual. This is fruitful, for at least two reasons. First, it allows many seemingly disparate semantic phenomena to be treated univocally; that is, it captures what is in common among many different kinds of linguistic constructions. Second, it allows language to be treated like a calculus, and that is one of the distinctive and fruitful ideas underlying the subsequent accomplishments of modern logic.

As to the first point, much of the beauty and power of Frege's function-argument analysis of language lies in its broad range of application: it is not just a more insightful way to break up a proposition than the subject-predicate distinction, it is also applicable to many sub-sentential phenomena. Many linguistic expressions behave, semantically, like functions. I will now survey the four different cases of operators (such as 'the father of ( )'), connectives (such as '( ) and ( )'), predicates (such as '( ) walks'), and quantified noun phrases (such as 'All ( ) ( )').

First, operators are expressions that behave semantically as functions from objects to objects. Common examples of mathematical operations include '( )^2' and '( ) + ( )'—the value of these operator-functions, given appropriate arguments, is a number. Operators in English include: 'the capital city of ( ),' 'the first child of ( ) and ( ).' Feed in an appropriate sort of argument—in the first case, a political region; in the second case, two people—and the function outputs the appropriate value for that argument—i.e., a city or a person, respectively. More generally, an n-place operator is a function from n individuals to an individual. It takes n individuals as argument and outputs an individual as value. The examples given above are all one- and two-place operators, but there is no logical or semantic limit to the number of argument-places an operator can have.

Second, the truth-functional connectives. Modern propositional logic, that most basic branch of modern logic, studies the inferential properties of the truth-functional connectives. A connective is a word which connects sentences together to make a bigger sentence, such as 'and' or 'because.' In this case, Frege's innovation is to view connectives as functions from sentences to sentences. Feed in two sentences as argument, and the connective-function outputs a sentence as value. A truth-functional connective is special kind of connective, the truth-value of whose output-sentence depends only on the truth-value of the input-sentences. Compare:

(1) World War 1 occurred and Princip shot Ferdinand.
(2) World War 1 occurred because Princip shot Ferdinand.

To know whether (1) expresses a truth, one need only know the truth-value of both constituent sentences: if World War 1 occurred, and if Princip shot Ferdinand, then (1) is true, and that is all there is to it. However, knowledge of the truth-values of those two constituent sentences alone does not suffice to allow us to know whether (2) is true. It is making a claim much stronger than (1); there is more at stake here than merely the truth-values of the constituent sentences. Clearly, and definitively, two people could agree on the truth-value of both constituents but yet disagree as to the truth-value of (2). So (2), as opposed to (1), is not a truth-functional claim; 'and' is, and 'because' is not, a truth-functional connective.

In addition to '( ) and ( ),' the one-place connective 'it is not the case that ( )' and the two-place connectives '( ) if and only if ( )' and '( ) or ( )' are other examples of truth-functional connectives. The former is a function from one sentence to a sentence, the latter is a function from two sentences to a sentence; and in all three cases, from a logical point of view, the truth-value of the output-sentence depends only on the truth-value of the input-sentence. The validity of many inferences depends on the logical properties of these truth-functional connectives; propositional logic studies such inferences by treating these connectives as functions from truth-values to truth-values. The truth-function denoted by '( ) and ( ),' for instance, is that function which yields 'true' as output when given the pair 'true, true' as input, and yields 'false' as output when given any other combination of values as input. Each one of the other truth-functional connectives mentioned above can be defined just as simply.

Propositional logic is the first and most fundamental department of modern logic. It is first developed in Frege (1879) *Conceptual Notation*, and then in a more accessible form by Russell and Whitehead (1910) *Principia Mathematica*. This brief introduction to it imparts a sense of the calculus-like approach to language and inference which Frege and Russell share, and which is characteristic of modern logic. One studies truth-functional connections among propositions by treating them as precisely defined functions from truth-values to truth-values. The resulting logical system has a much broader range than traditional logic: for one thing, it applies to any proposition whatever, not merely those which can be forced into the categorical mold; for another, it is not confined to syllogistic patterns of inference, but rather the number of truth-functional patterns of inference is infinite. The resulting logical system, in contrast to the rigid confines of the traditional categorical system, admits of exponential growth.

The third sub-sentential application of the function-argument distinction affords a distinctive and fruitful picture of what it is to make an assertion about something. Consider:

> Statements in general ... can be imagined to be split up into two parts; one complete in itself, the other in need of supplementation ... Thus, e.g., we split up the sentence
> *Caesar conquered Gaul*
> into 'Caesar' and 'conquered Gaul.' The second part is 'unsaturated'— it contains an empty place; only when this place is filled up with a proper name does a complete sense appear. ([5]: 151)

Something related to the subject-predicate distinction survives in Frege's work, in the case of object-dependent propositions. Even though that traditional distinction is dismissed from its role of central importance for logic, there remains a difference between the saturated things that can function as grammatical subjects and the unsaturated things that serve as predicates. In Frege's terms, the saturated argument-terms stand for 'objects,' and the unsaturated function-terms stand for 'concepts.' The meaning of a predicate is a concept; objects are the sorts of things that can serve as arguments for these functions. Concepts are special kinds of function—functions from objects to truth-values. For instance, the above '( ) conquered Gaul' is a function that yields 'true' for the argument 'Caesar' and 'false' for the argument 'Mother Theresa.' Concepts can be functions of two arguments (e.g., '( ) loves ( )'), of three arguments (e.g., '( ) put ( ) ( )'), and so on up. Again, there is no logical or semantic limit to the number of argument-places this kind of concept-function can have—although it is hard to see what use a 7-place concept would be, let alone a 17-place, or 700-place, concept.

So the meaning of a predicate is a concept, the meaning of a referring expression is an object. Concepts are functions that take objects as arguments and output truth-values. An n-place concept is a function from n objects to a truth-value. Herein lies a distinctively Fregean deep and insightful innovation: the essence of predication, the very logical core of what it is to say something about something, can be captured by a mathematical function. The nature of a function is to divide a range of objects into sub-groups, according to whether they have or lack some relevant property or properties. Frege sees that predicates in natural language—'( ) is walking,' '( ) loves ( ),' and so on—do exactly that, from a semantic point of view. In the case of an object-dependent proposition, one needs both a function and an argument in order to say something for which the question of truth or falsity arises. At the very limit of generality, what it is to make an utterance about something is to plug an argument into a function.

Once again the approach to language as a calculus is evident, and is seminal. This view of predicates standing for functions from individuals to truth-values is the core of the predicate calculus, the second strata of modern logic. The central idea of the predicate calculus is to treat the utterance that x is F as akin to the mathematical function $F(x)$. Object-dependent propositions consist of arguments and functions, and such propositions are true if and only if that object lies in the set of things which satisfy the predicate. There is, however, one more important ingredient required to give the full flavor of the predicate calculus, and it is our fourth and final sub-sentential application of the function-argument analysis.

This approach also affords the proper analysis of the semantics of quantified noun phrases. This is among the biggest steps forward here, the first and most enduring success of the project of philosophical logic. What, then, is the logical form of 'All Fs are G,' or of object-independent propositions more generally? Consider, for example:

(1) All giraffes are spotted.

The first step is to identify the function '( ) are spotted.' Next, what is the argument? An immediate candidate is the grammatical subject 'all giraffes,' but our arguments against categorical analysis, logged in the last section, sour this approach. That suggestion does not capture how (1) differs from object-dependent utterances, such as 'Jerome is spotted.' It misses the import of the fact that (1) expresses an object-independent proposition; it results in an inappropriate view of its inferential properties. There must be something saturating this function; for what we have here is clearly significant, clearly something for which the question of truth arises. So what could the argument be?

Frege looks to the variable, that vehicle of generality in mathematics. For instance, 'x + y = y + x' expresses that addition is commutative, with 'x' and 'y' standing for any two numbers whatever. Variables are employed semantically to express generality, and syntactically, to mark the gaps in mathematical functions. Given that predicates behave semantically like functions, they may well have a place here, in representing the logical form of general, object-independent propositions. His hypothesis is that the logical form of (1) is:

(1*) For any x, if x is F then x is G.

Here 'F' stands in for '( ) is a giraffe' and 'G' for '( ) is spotted.' What (1) says, essentially, is: take anything you like, it has the following truth-functional property—if it satisfies the predicate '( ) is a giraffe,' then it satisfies the predicate '( ) is spotted.' The variable ranges over all elements of the contextually salient domain.

The innovative idea here is that variables hold the key to understanding the semantics of quantified noun phrases. Whenever one has an object-independent proposition, there is a statement of generality, from a semantic point of view: all or some members of a given set are under discussion. Whenever you have an 'all' or 'some,' there is, syntactically, a gap to be filled in, a place to be marked, as opposed to there being some specific argument being plugged into the relevant function. Hence, in these cases, there is a variable at work, from a logical point of view, specifying the range of the predicate. (2*), for instance, gives the logical form of (2):

(2) Some giraffe is spotted.

(2*) There is at least one x such that x is F and x is G.

This takes us to the heart of why it is that object-independent propositions express relations among concepts, not assertions about individuals. There are no individuals in or around the content of such propositions, there are only variables, filling in the gaps in the predicates. These utterances are about concepts—about giraffehood and spottedness, in these particular cases—not about any specific objects.

Frege treats quantifiers as functions that take concepts as arguments and yield truth-values as output. 'All ( ) ( )' is a function which takes two concepts as argument and yields the value 'true' as output if and only if everything which the first concept is true of the second concept is true of as well; 'some ( ) ( )' is a function which takes two concepts as argument and yields the value 'true' if and only if there is at least one thing to which both concepts apply. Given the resources of propositional logic, other quantifiers are easily definable, such as 'No ( ) ( ),' 'Only ( ) ( ),' 'Exactly two ( ) ( ),' and so on. These are one and all functions from concepts to truth-values, whose value, in any particular case, depends only whether the relevant relation holds among the relevant concepts.

Frege's approach to generality, to the semantics of object-independent propositions, has subsequently become canonical. Consider a couple of cases from the last section, which were shown to exceed the means of categorical analysis.

(1) Every number is greater than some number.
(2) If a number is even, then the square of that number is even too.

Their logical forms, on Frege's view, are as follows:

(1*) For all numbers x, there is a number y, such that xRy.
(2*) For all numbers x, if x is F, then A(x) is F.

'R' stands for the two-place predicate '( ) is greater than ( ),' 'F' stands for '( ) is even,' and 'A( )' stands for the one-place operator 'the square of ( ).' Now consider:

(3) Everywhere you go, every boy is always in love with some girl.

In the last section, it was asserted that Frege's analysis accommodates utterances with as many quantifiers as you like, whereas categorical analysis only has the means to identify one quantifier per utterance. Here are two logical forms, which formalize two different readings of the ambiguous (3):

(3*) For all places x and all times y, for every boy z there is a girl w, such that, at x and y, zRw.

(3**) For all places x and all times y, there is a girl w such that for every boy z at x and y, zRw.

(3*) is the interpretation that takes (3) to be an utterance about all boys, always and everywhere, that they each love some girl or other; (3**) is the interpretation that takes (3) to assert that, always and everywhere, there exists a girl that all boys love. This approach to the logic of quantifiers allows one to break into the sentence and to formalize a wide range of relations among sub-sentential parts, in a way that exceeds by leaps and bounds the resources of traditional logic.

These are the three central branches of these first developments of modern logic: the logic of quantifiers, the predicate calculus, and propositional logic. The many and multi-faceted developments that have sprouted from this foundation are a testament to the power of this system. The development of this logic is itself just one of many testaments, outlined in this section, to the beauty and power of this function-argument analysis of language.

Most generally, this is how language works: start with a function-expression, plug an argument into the gap, and the result is a completed or saturated expression. Meaning itself is functional—all significant expressions are like the operation of addition, '( ) + ( ),' in that to grasp their meaning is to have the ability to ascertain the value, for a given argument. To know the meaning of '( ) + ( )' is to be able to identify a value for any appropriate arguments; the same holds for knowing the meaning of 'the capital city of ( ),' '( ) is happy,' 'All ( ) ( ),' and so on. The essence of a function is a general rule for identifying a value, given appropriate arguments; and competence with meaning can be helpfully characterized in these terms, as a grasp of a general rule for identifying the significance of an expression, whenever and wherever it occurs in an utterance.

This generalizeability attests to the fact that Frege has gotten to the bottom of something here. It is a truism that the reason why competent speakers can understand sentences they have never before encountered is that their meanings depend on the meaning of their parts and the way in which those parts are structured; it is Frege's insight that there is a structure in common among many different kinds of meaningful parts: the meaning of many molecular, structured linguistic expressions lies in the combination of these two fundamentally different kinds of expression. The traditional view was right to think that meaningfulness is the conjunction of two basic elements, but wrong to think that those elements are precisely the grammatical subjects and predicates. The two basic elements are rather functions and arguments. This gets to the bottom of the subject-predicate divide in a way that is sensitive to the crucial difference between quantified noun phrases and referring expressions, and that applies quite generally across a central and significant range of expressions and constructions. That range has been extended by subsequent work in philosophical logic, as many other kinds of natural language expressions have been shown to be illuminatingly analyzed via this function-argument approach to language.

IV.iii Analysis of Language as Philosophical Method

This section begins with an overview of Frege's and Russell's seminal contributions to the ideal language tradition, and then moves on to survey some of the things that sprouted from there, some philosophical morals and methods. The languages which the logicists designed in order to demonstrate that mathematics can be reduced to logic are extensional formal languages, in which there is an isomorphism between syntax and semantics. At least three concepts here need elucidation: formal, isomorphism, and extensional.

First, the language is called 'formal' in that it has no specific vocabulary, in that its sentences are not about anything in particular. 'Form' in this sense contrasts with 'content,' and there is a clear sense in which these logical languages lack content. Sentences in it consist of variables and symbols for logical functions and operations; they represent logical forms, not anything in particular. This is so for want of generality. Logic is about the most abstract features of a chain of reasoning, those things that it can share with many different arguments, concerning many different contents. Russell puts this point as follows:

> In logic, it is a waste of time to deal with inferences concerning particular cases: we deal throughout with completely general and purely formal impli-

cations, leaving it to other sciences to discover when the hypotheses are verified and when they are not. ([13]: 269)

(Perhaps even better: logic is a "subject in which we never know what we are talking about, nor whether what we are saying is true." ([10]: 221)) So the end here is a logical scaffolding of a language, a purely formal skeleton designed solely to test for formal validity.

Second, the languages are crafted so that there exists an isomorphism between syntax and semantics, which is to say that there is one precise semantic role systematically correlated with every linguistic expression. Every symbol corresponds unambiguously to a semantic bit, to something that must be reckoned with in the study of inference; and no two expressions do exactly the same thing, from a logical point of view. This notion of a tight one-to-one relation between syntactic expression and semantic role is a crucial concept in the ideal language tradition. It is a step toward Leibniz' dream of a *calculus rationator*, a language in which "... mere adherence to grammar would guarantee correctness of thought process." ([2]: 120)

This second feature is a response to the discrepancies between language and logic that engender the project of philosophical logic, a response to what Frege calls the 'logical defects' of language. Ambiguity, vagueness, and context-sensitivity are firmly outlawed in the conceptual notation, and so there is no possibility of being led astray in one's deductions by grammar. Note also that this notion of an isomorphism between syntax and semantics is an important step toward the development of the digital computer; i.e., insofar as a computer is a syntax-driven machine designed to perform semantic functions, these early formal languages are crucial and seminal steps toward its development.

Thirdly, these languages are extensional. The substance and significance of this notion will take a little more to explain. The term 'extension' contrasts with 'intension;' these terms refer to two dimensions or aspects of meaning. The extension of a term is that to which the term is correctly applied, the intension of a term is its significance or conceptual content. To illustrate, consider the difference between the names 'Hesperus' and 'Phosphorus' (the former was introduced to name the heavenly body known as the Evening Star, the latter for the heavenly body known as the Morning Star; as it turns out, the planet Venus is both the Evening Star and the Morning Star) or the predicates 'renate' and 'cordate' (the former means 'organism with kidneys,' the latter means 'organism with a heart'). Even though these pairs of terms agree in extension—in referring to the same object or objects—that does not entail that they are synonyms. 'Intension' names the dimension along which the meanings of such co-extensive pairs of terms can differ. There are several perennial debates in philosophy and logic about the significance and proper understanding of this intension-extension distinction.

A language is extensional if and only if expressions with the same extension can be interchanged without changing the truth-value of sentences in which they occur. On this function-argument approach to language, 'extension' refers to the output of the relevant function for the relevant argument. The extension of an argument-term is an individual, the extension of a predicate-term is the set of individuals that satisfy the predicate, and the extension of a sentence is a truth-value. The artificial languages designed for the purposes of modern logic are extensional in this sense: i.e., expressions—be they argument-terms, function-terms, or sentences—are treated as logically equivalent if and only if they have the same extension.

(A question here arises: If in these languages there is an isomorphism between syntax and semantics, then how can there be two different expressions that agree in extension? Either there is no one-to-one correlation between words and meanings, or else there are no two co-extensive expressions. Answer: 'the square root of 81' and 'the square of 3' are co-extensive, but do not constitute a counterexample to isomorphism, because they are built up out of different parts, each of which is perfectly unambiguous. As I explain in section V.i, Frege argues that this phenomena of co-extensive expressions which are not synonymous is pervasive.)

These logical languages are extensional: co-extensive terms are interchangeable, for all logical purposes. Take propositional logic, to start. In determining the truth-value of 'P and Q,' only the truth-values of the constituent sentences are relevant. If P and R have the same truth-value, then 'R and Q' is truth-functionally equivalent to 'P and Q.' Propositional logic, with its eye fixed on general patterns of inference, draws no further distinctions. Similarly for the case of predicate logic. The truth-value of 'x is F,' where x is the argument plugged into the function '( ) is F,' depends only on the extension of these terms, not on what they connote or evoke. Substituting another term 'y' with the same referent as 'x' should not have any effect on the inferential properties of the sentence—on what it entails and what entails it. The same holds for predicates: substituting another predicate that is true of all and only the same individuals should not result in any logically relevant differences. Thus, the argument-terms 'Hesperus' and 'Phosphorus' are treated as equivalent, for the purposes of logic—interchanging such pairs of names should not alter the truth-value of a sentence—as are the function-terms '( ) is a renate' and '( ) is a cordate'—they should output the value 'true' for all and only the same arguments. Finally, the same point holds for quantifiers. For instance, 'All ( ) is not ( )' is treated as logically equivalent to 'No ( ) ( ),' because they output the same truth-value for all and only the same pairs of concepts or predicates.

For a variety of reasons, many thinkers have found this property of extensionality to be desirable within and indeed beyond a system of logic. An extensional semantics is relatively clear and simple, because it avoids difficult and perhaps intractable questions about synonymy, about when two concepts should be taken to be intensionally equivalent. It avoids getting ensnared in debates about criteria of identity and distinctness for intensions, such as concepts or ideas or meanings or propositions, debates that do not admit of conclusive resolution. "Intensions are creatures of darkness," says Quine (1956: 178), "and I shall rejoice with the reader when they are exorcized." The cause of Quine's ire here is that intensions are not things about which we are capable of attaining the degree of precision desired in the systematic study of logic.

Wherever frustration with, or suspicion of, these abstract and shadowy intensions arises, or even just a desire to avoid getting ensnared in puzzles and conundrums concerning them, there is some reason to restrict oneself to extensional terms. Insofar as it is not the province of logic to tell us what concepts are, or when two words are identical in intension, it is entirely appropriate that these logical languages be extensional. Extensionalism is the orthodox position in modern logic; it studies inferences concerning individuals, sets, and truth-values, not meanings or concepts. (I should point out that intensional logics have been developed over the course of the twentieth century; and, further, these intensional logics have drawn inspiration from Frege, for reasons that will become evident in section V.i.)

These three properties are central to the logicists' contributions to the ideal language tradition. The aim is a language in which all and only differences in inferential properties affect content, a more rigorous instrument for making inferences. Thoughts expressed in these languages have a degree of precision that can only be approximated in natural language. The closer we can come to a transparent link between language and thought, to this isomorphism between medium and content, the more precise will be our reasoning.

In this way, these formal developments in modern logic blossom into the more ambitious and wide-ranging project of philosophical logic. Once it is recognized that logical form is not adequately captured by traditional categorical subject-predicate analysis, a whole new philosophical project is launched. Frege holds that not just Kant's view on mathematics, or the ontological argument for God's existence, but indeed much of traditional philosophy suffers from semantic sloppiness, and that philosophical logic can be applied to solve or dissolve many of the problems of philosophy. With this point Russell, Wittgenstein, and a host of other twentieth-century philosophers wholeheartedly agree. Russell consistently insists on the need for philosophical logic within the discipline of philosophy, because of the wealth of conundrums and problems which have resulted from inattention to semantics, and the many debates which continue to be intractable as a result of vagueness and imprecision in their conception and formulation. Consider also Wittgenstein (1921):

> 4.002 ... Language disguises thought. So much so, that from the outward form of the clothing it is impossible to infer the form of the thought beneath it, because the outward form of the clothing is not designed to reveal the form of the body, but for entirely different purposes.
>
> 4.003 Most of the problems and questions found in philosophical works are not false but nonsensical. Consequently we cannot give any answer to questions of this kind, but can only point out that they are nonsensical. Most of

the propositions and questions of philosophers arise from our failure to understand the logic of our language....

Thus develops the analytic approach to philosophy, with Frege as its great progenitor, Russell as its high priest, and Wittgenstein as both leading acolyte and fiercest critic. Analysis is the process of breaking down something into its constituent bits, so that its logical structure is displayed. This process becomes a meticulous methodology that is applied more generally across the whole discipline of philosophical questions and problems. Philosophical issues should be approached by breaking down complex problems into their simpler parts. Pay very careful attention to the way in which a problem is formulated, since many philosophical problems are the result of misleading formulations. Clarity is among the highest goods within this tradition: one of the first steps toward any philosophical problem is to formulate a problem in a way that reflects its logical structure.

The defining tenet of the analytic tradition is the use of philosophical logic as a means of identifing and formulating questions, and as a method in seeking to solve them. Philosophy is seen as an exercise in achieving clarity of thought by breaking down complex problems, claims and concepts into simpler constituent parts and logical relations. Thus again Wittgenstein (1921):

> 4.112 Philosophy aims at the logical clarification of thoughts. Philosophy is not a body of doctrine but an activity.... Without philosophy thoughts are, as it were, cloudy and indistinct: its task is to make them clear and to give them sharp boundaries.

This analytic movement has done philosophy a wealth of good. It has helped to clarify the confines and structure of many perennial philosophical claims, questions, and debates. It has spurred the development of some valuable tools and methods of philosophical analysis; it has increased our awareness of the importance of semantic acuity for any systematic inquiry; and it has raised the standards of rigor that characterize the discipline. However, as is perhaps the fate of all good ideas, some oversimplifications and overgeneralizations were perpetrated in the name of analytic philosophy. To close this section, let us investigate two closely related suspect directions in which analytic philosophy grew: one is toward the claim that philosophical logic is the only respectable task for the respectable philosopher; another is toward the claim that philosophical logic is the ultimate arbiter not merely in semantics but in metaphysics as well.

Frege and Russell, most specifically in the fields of semantics and mathematics, succeed in demonstrating that certain conundrums are caused by misleading formulations, and succeed in applying modern logic toward rectifying the situation. The early stages of analytic philosophy are built around certain very clear steps forward in a definitely circumscribed realm—the semantics of quantified noun phrases is perhaps the paradigm case, with Russell's theory of descriptions (to be discussed in section V.ii) also vying for that title. Many successors—indeed, Russell himself—were emboldened by these early successes. They believed that these cases demonstrated that all philosophical problems had this structure: careful semantic analysis would either afford a logical solution, or else would demonstrate that the problem was no real problem at all but rather the result of linguistic confusion. The suspect claim made by certain early analytic philosophers is that all real philosophy is philosophical logic, that anything which purports to be philosophy but is not philosophical logic is idle and worthless speculation. The paradigm here is surely Ayer's (1936) *Language, Truth, and Logic*.

One of the problems with this bold claim is that it is demonstrably false. Counterexamples abound: i.e., there is an immense tradition of philosophical works that are neither worthless speculation nor solely the analysis of the logical form of problems, and new chapters and volumes are added all the time. Apart from this damning *a posteriori* evidence, there are also rather large *a priori* problems. Perhaps chief among them is that the claim that philosophical logic is the only worthwhile task of the philosopher is not itself identified or justified via philosophical logic: the very claim itself contravenes its own strictures. The upshot is that these strictures are woefully unreasonable. Philosophical logic is important, but, *per se*, it is blind. Other sorts of philosophical inquiry are required to identify and justify the ends to which philosophical logic is to be put.

Second, let us survey some putative metaphysical implications of philosophical logic. As Frege conceives it, the goal of philosophical logic is to lay bare the structure of thought. Given certain developments explained in the next couple of sections, Russell and Wittgenstein were to replace the thought with the fact as the basic semantic unit. That is, whereas Frege takes the content expressed with an utterance to be a thought, Russell and Wittgenstein take the content expressed with an utterance to be a fact. With that shift comes the claim that the analysis of language is capable of laying bare the structure of facts, of reality itself. This claim that philosophical logic reveals the basic structure of what there is in the world defined a research project spearheaded by Russell and Wittgenstein, and which influenced many succeeding philosophers. (This research program is reminiscent of Aristotle's *Categories*, and of some of the medieval work on the relation between language and reality which it influenced. The distinctive features of Russell's and Wittgenstein's project concern the use of the powerful new tools of modern logic.)

Russell (1918) and Wittgenstein (1921) hold that, in their analysis of language, they are uncovering the basic metaphysical simples, the logical atoms out of which reality is composed. Frege's thought-expression isomorphism becomes an isomorphism between language and reality, the result being a line of thought along these lines:

P1: In an ideal language, the syntax of the language mirrors the structure of the facts described or depicted by language.
P2: The language of modern logic is an ideal language.

C: Therefore, reality itself, the mind- and language-independent world, has the structure depicted or captured by this ideal language.

Analysis of language becomes the method in metaphysics: the world itself has the structure prescribed by modern logic. Reality consists of arguments coupling with functions to produce facts. As the above conclusion intimates, Frege's syntax-semantics isomorphism becomes a picturing relation in the hands of his successors: a certain sentence S expresses or refers to a certain fact F, in virtue of its depicting, being a picture of, that fact. The sentence 'Jane is to the left of Bill' is a picture of a certain fact; sentences that do not yet look like pictures of facts just need elucidation via philosophical logic. Either they can be demonstrated to be such, or they are meaningless. As Urmson observes: "The aim of analysis thus was to make every statement an adequate picture of the reality it referred to, and the perfect language was the tool which could make the undertaking capable of complete realization." (1956: 21)

It must be conceded that the doctrine of picturing has a very limited range of application. It is most plausible in the case of relational propositions—such as 'Jane is to the left of Bill,' whose logical form is 'xRy.' The case for the claim that sentences picture facts is significantly less plausible for propositions of the form 'x is F,' such as 'John is tall.' The correlative fact is as respectable a fact as there is, but one really has to use one's powers of imagination to believe that that sentence *pictures* the relevant fact. When one gets to object-independent propositions one really loses much semblance of plausibility; not only is there the problem of explaining in precisely what sense does 'Humans are mortal' picture anything; there is also the problem of describing what fact it pictures. That would be a strange looking fact indeed, quite distant from the ordinary medium-sized empirically accessible facts these analysts take as their paradigm cases. Further, the problem of finding a negative general fact for 'There is no hippopotamus in this room' to picture was, at times, downright embarrassing.

For some time Russell wrestled with such things as negative facts and general facts. He conceded their existence because he needed them in order to render this view comprehensive; but he was never comfortable with them, and he came to reject this entire picture of meaningfulness. Wittgenstein (1953), too, came to criticize this approach as premised upon a terribly oversimplified conception of the nature of language. Describing facts is but one of the wide range of things that we use language to do, and building a theory of meaning entirely on this one kind of usage is really quite narrow, and engenders serious oversimplifications. Nonetheless, despite these rejections, this method of taking philosophical logic to be the ultimate arbiter in metaphysics is not without enduring influence. Consider, for instance, Quine's: "The doctrine is that all traits of reality worthy of the name can be set down in this austere idiom [i.e., the language of modern logic] if in any." (1960: 228) Quine takes extensional truth-functional logic as his criterion for what there is, and Quine is among the most influential metaphysicians of the twentieth century.

Whatever one thinks about that extreme position, philosophical logic still plays a role in metaphysical debates, and surely rightly so. Not many philosphers today would defend Ayer's claim that all philosophy is philosophical logic, or Quine's claim that anything that cannot be described in the austere idiom of modern logic thereby does not exist; but more or less all philosophical positions, in theory and in practice, accord with weaker versions of those maxims. In many ways, philosophical logic has changed the standards of the discipline for the better.

V. Some Disputed Issues in Early Analytic Philosophy

*V.i Meaning: Semantic Monism
and Semantic Dualism*

Traditionally, although questions about meaning were rarely addressed explicitly, when pressed as to what constitutes meaningfulness, philosophers were prone to say one of two things (or perhaps to answer with some blend of the two views, as each is well-suited to some proper subset of the expressions of natural language). These two views are psychologism—which we have met previously, in section II.i—and referentialism.

Referentialism is so-called because it identifies the meaning of an expression with that to which it refers. There are passages in the dialogues of Plato that clearly indicate his allegiance to the referentialist view; it is an ancient and eminently sensible view. It is a truism that we use language to express information about things in the world; referentialism is the view that meaningfulness consists of pairings of individual expressions with chunks of the world. Referentialism is best suited to proper names, such as 'Bertrand Russell' or 'Piccadilly Circus;' it is somewhat less well-suited to, but not implausible for, general or kind terms, such as 'tiger,' 'milk,' or 'Italian politician.' On this view, names like 'John' mean objects, predicates like 'is tall' mean sets or properties, and the meaning of a sentence such as 'John is tall' is the fact or state of affairs that the sentence is about.

Psychologism is the view that the meaning of an utterance is constituted by the speakers' images or ideas. The psychologistic view takes as primitive and fundamental the fact that, whenever an utterance occurs, a speaker is trying to communicate something; on this approach, it is this something that the speaker intends to communicate, the message that the speaker has in mind, that constitutes the meaning of the utterance. Like referentialism, psychologism is an intuitive and perennially prevalent view—both of these views seem to get something right, or, perhaps better, each is addressed to an indisputably important fact about, or aspect of, linguistic use. As explained in section II.i, psychologism was prevalent among Frege's and Russell's idealist and empiricist contemporaries.

Herein lies another central reason why Frege has such a seminal place in the history of the study of language: he conceives of, and articulates, arguments against both of these entrenched—some might have said jointly exhaustive—

views about meaning. Explaining the content and import of these arguments is the primary job of this present section, and it will proceed via three interrelated steps. First, I will describe Frege's arguments against these two views. Second, I will further investigate the distinction between sense and reference, the core of Frege's argument against referentialism. Finally, I will explore some questions about the semantic dualist position that results from Frege's positing of senses.

Some of the central arguments against the psychologistic approach were outlined in section II.i. On this question Frege sets in motion a snowball, which gained momentum as more and more people thought about these questions about meaning, and which many believe to have been fashioned into a knock-down argument by Wittgenstein (1953). One fundamental problem is that an idea is private to an individual speaker, whereas meaning is the common property of a group of speakers. Another is that most linguistic expressions—'of' or 'justice,' for instance—just simply do not signify any definite idea or mental image. Psychologism does not give due gravity to the point that we talk about individuals and states of affairs, not about our ideas of them: when I tell you that it is raining outside, or that there is trouble at the henhouse, I am talking about something in the world, not about something inside my head.

These arguments against psychologism are seminal historically in establishing the philosophy of language as an autonomous area of inquiry because they show that the study of meaning is not a province within psychology: linguistic meaning cannot be isolated inside the head of any individual speaker. If we are to study meaning systematically, we must keep clear the distinction between it and the unlimited and various subjective associations and ideas which speakers privately attach to linguistic expressions. A line of criticism that applies to both psychologism and referentialism is relevant here. It is Frege's context principle: "Never ask for the meaning of a word in isolation, but only in the context of a proposition." ([4]: 140) This is a departure from these traditional approaches to meaning: the demand that each individual expression have a meaning in isolation commits one to finding some entity to serve as its meaning, and so has propelled theorists toward psychologism, or referentialism, or some blend of the two. In contrast, Frege's function-argument approach takes the sentence to be the basic semantic unit, methodologically and ontologically. The meaning of an individual expression is the contribution it makes to propositions. To set things up in these terms undermines the impulse to find some discrete entity to serve as the meaning of every meaningful expression. This methodological point, and its ontological consequences, has been very influential in subsequent thinking about meaning.

Frege's arguments against referentialism constitute the second of the three central distinctions at the bottom of the philosophy of language, along with his ([1], 110-11) distinction between quantified noun phrases and referring expressions and Russell's ([11], 235) distinction between referring and denoting (the subject of the next section). The key premise of the argument is that there are expressions with the same referent but which differ in meaning, that there can be an objective semantic difference between co-referential expressions. Thinking about identity statements precipitated this argument. Frege ([7]) begins: Is identity a relation between objects, or between signs that signify objects? He proceeds to argue that neither of these can be the case. Consider the following:

(1) Holland is the Netherlands.
(2) Hesperus is Phosphorus.
(3) David Berkowitz is the Son of Sam.

((3) is one example of the sort of case in which a name is introduced to refer to the perpetrator of a series of crimes before the identity of the criminal is known. 'Jack the Ripper' is probably the most familiar example to English speakers; 'the Unibomber' is a more contemporary case. 'Son of Sam' denotes the person responsible for a series of murders in New York City in the mid-1970s. David Berkowitz has been convicted of those crimes.)

Frege originally held a referential view of meaningfulness, according to which the semantic role of a name is, simply and wholly, to introduce an argument into a function. Right from the beginning ([1]: 108-09), though, he recognizes that identity statements pose a problem for this view, because of the difference between (1)-(3) and:

(1a) Holland is Holland.
(2a) Hesperus is Hesperus.
(3a) David Berkowitz is David Berkowitz.

Frege sees that the referential view entails that these pairs of sentences express the same proposition—exactly the same function, exactly the same arguments. This is unacceptable, says Frege, because there is clearly a semantic difference between these two types of utterance. (1a)-(3a) are analytically true and knowable *a priori*—they assert that a certain individual is self-identical, and that is trivially true and uninformative. (1)-(3), in contrast, are neither true in virtue of meaning nor knowable independently of experience, according to Frege. One can know how to use the expressions 'Hesperus' and 'Phosphorus,' 'David Berkowitz' and 'the Son of Sam,' and do all the armchair theorizing and logical deductions which can be done with respect to those terms, but yet still not know whether (2) or (3) express truths.

So the informativeness of some identity statements rules out the first option cited above, that identity is a relation between objects. To say that identity is a relation between signs for objects is to say that (1)-(3) are about labels, are about how words are used. Frege ([1]: 109) originally held such a view. Essentially, the suggestion is that names perform a dual semantic role only in the case of identity statements—they constitute a unique complication to the referential view, the sole case where a name stands for itself, in addition to performing its typical role of standing for its referent. The two names in (1)-(3) both introduce an argument into a function and stand for themselves; and this accounts for the difference in informativeness between them and (1a)-(3a).

There are problems with this view that identity is a relation between signs of objects. First, it does not afford a satisfactory account of the informativeness of certain identities, of the fact that statements like (1)-(3) do convey substantive information about how things are in the world. The original recognition of the truth of (2) and (3), for instance, are significant extensions of human knowledge. However, this view assimilates such utterances to "I have decided to change my name to 'Larry,'" or to "Mrs. Smith's maiden name is 'Jones,'" which do not *per se* significantly enrich one's view of how the world is. A key flaw of this view is that it takes identity statements to be about words, whereas, as those examples along the lines of (2) and (3) illustrate, identity statements can be far from such trivial linguistic niceties. This view renders identity statements rather thin and arbitrary, whereas they seem to be capable of constituting substantive extensions of knowledge.

Identity is as simple and transparent a function as there is: it outputs the value true if and only if one inputs as argument two terms which denote the same entity. So there is something unsatisfactory about a view that attributes mysterious properties to the identity predicate. It seems not just *ad hoc*, but wrong. Surely, the following is not just a decision about how signs ought to be used:

(4) the square root of 81 = the square of 3.

It rather gets at a truth about relations between certain mathematical objects. Even more surely, it differs in informativeness from:

(4a) 9 = 9.

These are among the considerations which push Frege away from the view that informative identities show something is queer about the identity predicate, and towards the view that they show that something is wrong with the referential view.

Ultimately, though, the consideration that knocks down this view that identity is a relation between signs is that this difference in informativeness among co-referential expressions generalizes. Frege holds that the following pairs can differ in truth-value:

(5) The FBI is looking for the Son of Sam.

(5a) The FBI is looking for David Berkowitz.

(6) Joe believes that Hesperus is the brightest light in the evening sky.

(6a) Joe believes that Phosphorus is the brightest light in the evening sky.

This generalizeability all but proves that the problem concerns not the identity predicate, but rather the notion of information itself. Some third account of the semantics of identity is required. Referentialism does not need a band-aid. It needs to be replaced.

We are now in a position to survey the other two central points of this section: the distinction between sense and reference, and the notion of semantic dualism. Frege concludes that co-referential expressions can fail to be synonymous, that there can be an objective semantic difference between expressions that refer to the same thing. It follows that meaning cannot be identified with referent, and so we must look elsewhere for what constitutes the meaning of an expression. He introduces the term 'sense' to denote the dimension of meaningfulness along which such pairs of expressions can differ. Statements like (1)-(3) are informative because they express that two expressions having the same reference can differ in sense.

The sense-reference distinction is perhaps best illustrated in the case of a definite description, such as 'the tallest woman in Canada' or 'the architect who designed this building.' The sense of these expressions is the uniquely identifying condition that they mean, that which must be grasped in order to understand these expressions. The sense is accessible to any English speaker, regardless of whether they have ever been to Canada or seen this building. The reference, in contrast, is the thing that the description picks out: i.e., one specific Canadian woman, one specific architect. The sense is an abstract conceptual condition that is of or about something, that is criterially linked to understanding the expression, and that is accessible to competent speakers, regardless of whether or not they have encountered the reference. The reference is the thing in the world that the sense picks out.

In general terms, the distinction between sense and reference is a distinction between what we say and what we say it about. Frege refers to sense as the mode of presentation of the reference, the way in which the reference is presented to the speaker. 'Holland' and 'the Netherlands,' or 'David Berkowitz' and 'the Son of Sam,' are two different linguistic means of presenting the same object. Sense constitutes what is said, communicated, the content expressed with an utterance. On Frege's view, the content of our thought and talk is made up of these abstract conceptual conditions, not the things themselves. The information expressed with an utterance—that which we are calling a proposition, but Frege calls a 'thought'—is entirely composed of senses.

Generalizing from the case of identity statements, any difference in informativeness is taken by Frege to be sufficient proof of a difference in sense. The term 'cognitive significance' is used to denote these differences of informativeness. Two utterances differ in cognitive significance if a competent speaker can take contrastive attitudes toward them (e.g., believe that one is true while withholding judgement on the second). Both Frege and Russell appeal to cognitive significance as their criterion of propositional identity—i.e., if two utterances differ in cognitive significance, then they express distinct propositions. Not only (2) and (2a), but also (7) and (7a), can differ in cognitive significance:

(2) Hesperus is Phosphorus.
(2a) Hesperus is Hesperus.
(7) Hesperus is the brightest light in the evening sky.
(7a) Phosphorus is the brightest light in the evening sky.

Hence utterances of these sentences express distinct propositions.

To recap: the sense of an expression is the abstract conceptual condition that it means. 'Renate'-'cordate' and 'Hesperus'-'Phosphorus' are examples of expressions that have the same reference but differ in sense. These terms make distinct contributions to propositional content. The proposition expressed by a sentence is a function of the senses of its parts and the way in which they are combined. If two sentences differ in cognitive significance, then they must express distinct propositions; if all of their parts are co-referential—as in the case of (2)-(2a) and (7)-(7a)—then at least one pair of expressions must differ in sense.

The sense of a sentence is a proposition; its reference is a truth-value. Frege explains this somewhat hastily in ([5]: 149ff), and gives some more careful arguments for it in ([7]: 179-82). Since every linguistic expression has both a sense and a reference, so must a sentence. Whatever the reference of a sentence is, it must be a function of the references of its parts. It must remain constant through substitution of co-referential parts. Frege holds that truth-values are well-suited to fit the bill. When you are enjoying a work of fiction, your interaction with the work is confined to the level of sense. To look around in the world to try to isolate the person who best fits the description of Sherlock Holmes, for instance, in order to determine the truth-value of some of the claims made in the book, is to misunderstand fundamentally the genre of fiction. Propositions about Holmes are significant and entertaining; but they lack a truth-value, in the strict sense, because 'Holmes' lacks an actual, flesh-and-blood referent. This, too, suggests to Frege the aptness of taking the reference of a sentence to be a truth-value. Whatever the reference of a sentence is, it must be systematically affected when a part of the sentence lacks a reference; and, again, truth-values fit the bill. When you wonder as to the truth-value of a sentence, you thereby become interested in the references of its parts, in how things stand with respect to the real-world correlates of the constituent parts of the sentence. To make a judgement, for Frege, is precisely to move from the sense of a sentence—the proposition expressed—to its reference—to the truth-value of the claim it makes.

Two comments, before moving on. First, there are connections between Frege's sense and the traditional notion of intension. This is a key reason why Frege has been a source of inspiration for intensional logicians, despite the extensional character of his own logic—many take Frege's arguments against referentialism to establish not just the respectability, but also the need, for a logic of senses or intensions in a comprehensive theory of inference. However, it must be pointed out that the connection between Frege's sense-reference distinction and the traditional intension-extension distinction is a matter of some dispute. Frege never appeals to that traditional distinction in arguing for or defining his distinction between sense and reference. Second, there is a tension between two dimensions or aspects of sense. There is on the one hand the communal aspect (i.e., sense is that which must be grasped in understanding the expression, accessible in common across a community of speakers) and on the other hand an individualist aspect, which is illustrated by the role that cognitive significance plays in individuating propositions—note well that cognitive significance is defined with reference to an individual speaker, not a community of speakers. Precisely what does understanding 'Hesperus' or 'Aristotle' or 'water' consist in? Is there one community-wide answer to that question? It is much easier to divide off the intersubjective semantic content from the private subjective associations in the case of 'the tallest woman in Canada' or 'the architect who designed this building;' it is notoriously difficult to do so—to identify one uniquely identifying condition, grasp of which constitutes competence with the expression—for the above referring expressions.

Frege himself struggles with this point—cf. ([7]: 176, n.2), ([9]: 207-08). Nevertheless, he helps himself to publicly accessible senses, out there and the same for all. He assumes that membership in a linguistic community is a matter of associating the same senses with the same linguistic expressions. This assumption looks hopelessly uncritical, a century later, especially in the case of proper names and natural kind terms—cf., e.g., Kripke (1972), Putnam (1975).

Now onto the third and final point in this section. The terms 'semantic monism' and 'semantic dualism' denote two styles of linguistic analysis within the Semantic Tradition. Semantic dualism is the view that there are two sorts of entity systematically correlated with every significant linguistic expression, that the content of what we say is distinct from the objects we say it about. Semantic monists, in contrast, believe that abstracta such as Frege's senses create more problems than they solve, so that semantics is better off without them. The content of our thought and talk is precisely the objects, stuff, and states of affairs at which those thoughts and utterances are directed. This is an issue which divides Frege from Russell, and which endures in the philosophy of language. Monists and dualists agree on such methodological maxims as Frege's Context Principle, agree that psychologism can not afford a satisfactory account of meaning, and that informative identity statements refute the naïve referential view. However, they disagree on the basic metaphysical question about what meanings are. What sorts of things are propositions? Are they made up of abstract or concrete entities? Does 'John' express a meaning that singles out John? Or does it just directly mean John?

The classic illustration of the difference between semantic monism and semantic dualism comes in correspondence between Frege and Russell in 1902. Frege detects what he takes to be a slip on Russell's part and corrects him: "... Mont Blanc with its snowfields is not itself a component part of the [proposition] that Mont Blanc is more than 4000 meters high." (Frege 1980: 163) Russell replies: "I believe that in spite of all its snowfields Mont Blanc is itself a component of what is actually asserted in the proposition...." (Frege 1980: 169) For Frege, propositions consist of abstract entities, senses, which are about individuals and states of affairs. For Russell, propositions literally contain those individuals and states of affairs.

Frege is the most influential semantic dualist. His argument that we must associate both a sense and a reference with every significant linguistic expression is the paradigm argument for countenancing meanings as distinct from things meant. Perhaps the biggest problem with semantic dualism is metaphysical: Exactly what, then, are these vague and shadowy, neither physical nor psychological, senses or meanings? The negative side of Frege's view—i.e., his arguments about what sense cannot be—is clear, poignant, and insightful. However, the same cannot be said of the positive side of Frege's view—i.e., of his answer to the question of what senses are, exactly. Frege certainly uses obscure and regrettable metaphors in elucidating his position: he takes his arguments against psychologism to have shown that senses cannot exist in Descartes' mental realm, the arguments against referentialism to have shown that senses cannot exist in Descartes' physical realm, and so posits some third realm in which senses can live and breathe. Here we see obscurities intersecting, as tendentious Platonist metaphors are coupled with the regrettable realm-talk of modern philosophy. This does little to explain what senses or meanings are.

Regardless of where one stands on the question of whether Frege's third realm is an instructive metaphor or an unreasonable metaphysical burden, though, Frege's challenge lives on, and that despite various ingenious attempts to stuff meaning back into the head or out into the world. Whatever the merits of Frege's positive views, we owe him a debt for showing us that meaning is a more complicated phenomenon than had previously been thought, that meaning cannot obviously consist in subjective psychological associations or in straightforward correlations between bits of language and bits of the world.

Russell and Wittgenstein develop semantic monist views, in which concrete facts play the role that abstract thoughts do for Frege. Before moving on to that story, I must say a little more about Meinong. As mentioned in section III.iii, Meinong is a key influence on Russell's semantics. Some grasp of his view is necessary in order to comprehend the import of Russell's theory of descriptions, the subject of the next section. Meinong was a follower of Brentano, whose lasting contribution to philosophy is his thesis that intentionality—i.e., being of or about something—is a distinctive mark of mental phenomena. (See Brentano (1884).) Meinong accepts the premise that all mental states are essentially directed at something. However, and contra Brentano's psychologism, Meinong draws the metaphysical conclusion that therefore there is something that every mental state is of or about. Meinong, in effect, existentially generalizes Brentano's: every mental state is directed at some object, to infer the conclusion: for every mental state, there is an object at which that mental state is directed.

Meinong's is a naive referentialist view: he holds that the meaningfulness of an expression entails that there be some thing that the expression means. For instance, much thought and talk has involved the expression 'the perpetual motion machine,' and therefore there must be some thing that that expression means; the very fact that I can describe to you the house I once intended to build but did not and that we can discuss it shows that, in some sense, there is some object about which we are talking; and so on. Meinong's referentialism is somewhat sophisticated, in that he distinguishes between actual flesh-and-blood existence and mere subsistence. On his view, the set of objects is divided into those that actually exist and those that merely subsist. Natural scientists, as well as most philosophers, have been most interested in the actual, to the exclusion of the non-actual objects, and understandably so. However, a comprehensive theory of the objects at which our mental states are directed, of the content of our thought and talk, must also embrace and accommodate non-actual objects.

From Russell through Quine and beyond, philosophers have criticized Meinong's metaphysics as bloated and unlovely, as too much for an honest philosopher to bear. They hold that, like Frege's senses (or perhaps even worse), non-actual subsistent objects create more problems than they solve. Regardless of where one stands on the metaphysics of possibilia, though, it is for semantic needs that Meinong countenances these non-actual objects. Meinong is a referentialist in need of a course in philosophical logic. One would be hard-pressed to find a more blatant instance of what Russell ([13]) criticizes as the subject-predicate fallacy, of the mistaken and pernicious view that every significant linguistic expression must mean some specific thing. Frege ([1], 109ff) does enough to render that view obsolete, but it would still be some decades before Frege's views were widely appreciated.

For some years, Meinong had a strong influence on Russell. Consider Russell's:

> *Being* is that which belongs to every conceivable term, to every possible object of thought—in short, to everything that can possibly occur in a proposition, true or false, and to all such propositions themselves.... "A is not" must always be either false or meaningless. For if A were nothing, it could not be said not to be; "A is not" implies that there is a term A whose being is denied, and hence that A is. Thus unless "A is not" be an empty sound, it must be false—whatever A may be, it certainly is. Numbers, the Homeric gods, relations, chimeras, and four-dimensional spaces all have being, for if they were not entities of a certain kind, we could make no propositions about them. (Russell 1903: 449)

Our next development concerns Russell's gradual move away from this realist-referentialist semantics. The theory of descriptions is his attempt to avoid Meinong's metaphysics without rejecting monism, to account for Frege's data without embracing senses. The strategy is to limit the range of referentialism, by drawing Frege's distinction between quantified noun phrases and referring expressions in a different place. According to Russell, quantified noun phrases are much more prevalent than Frege suspected.

V.ii Logical Form: On Denoting

In 1900 Russell wrote to Moore:

> Have you ever considered the meaning of *any*? I find it to be the fundamental problem for mathematical philosophy. E.g., 'Any number is less by one than another number.' Here *any number* cannot be a new concept, distinct from the particular numbers, for only these fulfill the above proposition. But

can *any number* be an infinite disjunction? And, if so, what is the ground of the proposition? The problem is the general one as to what is meant by any member of any defined class. I have tried many theories without success. (Coffa 1991: 103)

Russell is here in search of the semantics of quantification: the second last sentence, for instance, more or less defines the semantic role of the variable. Whereas Frege's rejection of naïve referentialism is precipitated by thinking about the semantics of identity, in Russell's case this role was played by unsuccessful attempts to find some thing for 'a' or 'any' or 'all' to name. For some time, in the early years of the twentieth century, Russell convinced himself that quantified noun phrases—'denoting concepts,' as he called them—were mysteriously and uniquely contrary to his otherwise uniformly referentialist semantics. If you see 'John' or 'justice' in a sentence, you can be sure that the proposition expressed is about John or justice. In contrast:

If I say "I met a man," the proposition is not about *a man*: this is a concept that does not walk the streets, but lives in the shadowy limbo of the logic-books. What I met was a thing, not a concept, an actual man with a tailor and a bank account or a public-house and a drunken wife. (Russell 1903: 53).

However, Russell was never comfortable with this intermediate view of denoting concepts. It is obscure and seems *ad hoc*; ultimately it stands out as a label for, rather than a solution to, a difficulty. That difficulty is: What is it about the semantics of 'any number,' or 'a man,' which sets it apart from the vast range of subject-expressions in natural language? What other expressions belong in this special category? By 1905 Russell has his solution together. He had rediscovered the semantics of quantified noun phrases.

So this question about the semantics of 'any number' and 'a man' is for Russell what informative identities are to Frege, in the following senses: First and foremost, it is that which shatters a rosy and uniform, naïve referentialist semantics. Both first tried to content themselves with a band-aid solution for some time (i.e., Frege's conjecture that identity statements express relations between signs; Russell's first theory of denoting). Both came to throw away the band-aid, in Frege's case by a principled rejection of referentialism, in Russell's case by a principled restriction of referentialism. Alternatively put, at least in part because Frege and Russell arrive at the semantics of quantification from divergent paths, they differ in their conception of exactly what should count as a quantified noun phrase.

Russell's semantics of quantification is identical to Frege's, in that quantifiers behave semantically as functions from predicates to truth-values. However, definite descriptions, phrases of the form 'the F,' are counted as referring expressions by Frege, but as quantified noun phrases by Russell. This theory of descriptions is both one of the most important differences between the views of Frege and Russell, and one of the most important ideas or developments in the philosophy of language. In keeping with Russell's ([11]) approach, I will state the theory of descriptions first, and then give some reasons for it.

The crux of the theory is the claim that the logical form of utterances of the form 'The F is G' is given by three quantifier statements:

(1) There is at least one F.
(2) There is at most one F.
(3) All Fs are G.

These are (1) the existence clause, (2) the uniqueness clause, and (3) the attribution of the predicate. (1) plus (2) account for the sense in which phrases of the form 'the F' denote the unique F, or exactly one individual that satisfies the relevant predicate. Given that there exists a unique F, (3) gives an object-independent semantics of the attribution of the predicate. This object-independent semantics is a distinctive feature, and a key virtue, of Russell's theory: even within a monist framework, the meaningfulness of 'The F is G' does not depend on there in fact being a unique F.

To illustrate, consider the example, 'The candidate who gets the most votes wins the election.' It is broken up into the following three statements:

(1a) At least one argument satisfies the predicate '( ) is the candidate who gets the most votes.'
(2a) At most one argument satisfies the predicate '( ) is the candidate who gets the most votes.'
(3a) For all things x, if x satisfies the predicate '( ) is the candidate who gets the most votes,' then x satisfies the predicate '( ) is the candidate who wins the election.'

The same is true of 'The perpetual motion machine would be a handy kitchen appliance.' It is meaningful, despite the fact that there is no such object. The utterance incurs no metaphysical commitments that 'a perpetual motion machine' or 'all perpetual motion machines' does not also incur, on a Fregean analysis; i.e., it commits only to the properties of perpetuity, motion, and machine-hood. This commitment to properties, not objects, makes all the difference to Russell: non-existent objects offend his 'robust sense of reality' ([14]: 281), but he is already and independently committed to realism about properties, in order to account for *a priori* knowledge concerning them—cf. (1912). Indeed, Russell's theory of descriptions can and has been combined with a wide range of different views on the metaphysics of properties.

This, according to Russell, gives the correct semantics of expressions of the form 'the F.' They are quantified noun phrases, not referring expressions. Propositions expressed by sentences in which they occur are object-independent, not object-dependent—what originally appeared to be reference to an object is cashed out in terms of quantifiers and variables. For this reason Russell calls descriptions 'incomplete symbols defined in use,' echoing what Frege says about quantified noun phrases at ([1]: 110-11): they do not designate any entity *per se*, but nonetheless make significant contributions to propositional content. To know the meaning of 'all Fs' or 'the F' is not to have some definite image in mind, or to be acquainted with some specific object; it is rather to be able to grasp the truth-conditions of any sentence in which those expressions occur.

Given the different semantic mechanisms that Frege ([1]: 110-11) sees underlying the object-dependent/object-independent distinction, Russell sees that the question of which of these categories a proposition belongs in is not a question about the number or amount of entities that the proposition is about. It is rather a question about the semantic mechanism at work. So a proposition can be about one individual, and still be object-independent, provided that the semantics of the designator is like 'every even number' and unlike '20.' This is precisely Russell's claim, about expressions like 'the even number between 10 and 13.'

Russell uses philosophical logic to build his case: the data concern the inferential properties of certain kinds of utterance which pose problems for some alternative views. According to the lore, the *Waverley* novels were published anonymously, and rumors circulated around Edinburgh that Sir Walter Scott was their author. As it turns out, these rumors were true. George IV is said to have wondered whether Scott is the author of *Waverley*. What, says Russell, is the content of this wonder? It must be different from the question of whether Scott is Scott, for "an interest in the law of identity can hardly be attributed to the first gentleman of Europe." ([11]: 240) This is remarkably similar to Frege's problem of informative identities. For Russell, the conclusion is that 'the author of *Waverley*' is a quantified noun phrase, not a referring expression; otherwise there is no accounting for the informativeness of the utterance that Scott is in fact that author, no accounting for what Frege calls the actual knowledge this conveys.

Note that this inference depends on the assumption of semantic monism, on a referentialist view of referring expressions: "'I think, therefore I am' is no more evident than 'I am the subject of a proposition, therefore I am' ..." ([11]: 240) (Russell does attempt to argue against Frege's semantic dualism, on which more in a moment.) Given semantic monism, plus the informativeness of the statement that Scott is the author of *Waverley*, it follows that that utterance expresses an object-independent proposition. It follows that at least one of the expressions is not a referring expression, and Russell infers that the non-referring culprit is 'the author of *Waverley*.'

Further, as in the case of Frege's informative identities, the essentials of this point generalize. The following can differ in cognitive significance, as one could clearly assent to (2) without assenting to (1):

(1) Scott wrote Waverley.

(2) The author of *Waverley* wrote *Waverley*.

Russell agrees with Frege that a difference in cognitive significance is sufficient proof of an objective semantic difference. Substituting a referring expression for 'the F' changes the meaning of sentences in which these expressions occur, even if those expressions have the same reference. Given the assumption that the meaning of a name is its referent, it follows that the meaning of a definite description cannot be its referent.

Or consider the difference between 'The perpetual motion machine does not exist' and 'x does not exist,' where 'x' is a name. The latter, Russell holds, is either false or meaningless: either the name refers to something, and so it is false to say that that referent does not exist, or else it is not really a name, in which case there is no statement made. However, 'The perpetual motion machine does not exist' is both perfectly meaningful and true. Therefore, he concludes, descriptions and referring expressions do not belong in the same semantic category.

Another puzzle which Russell uses to explain the virtues of his theory concerns the law of excluded middle, which Russell interprets to entail that either 'A is B' or 'A is not B' must express truths. However, when we substitute a description that does not denote for 'A,' we have a problem. It seems that neither of the following express truths:

(3) The present King of France is bald.

(4) The present King of France is not bald.

Neither the set of bald things nor the set of non-bald things numbers the King of France among its members. Russell's theory handles this problem nicely. (This is another of the early and paradigmatic successes of philosophical logic, of careful attention to semantic matters dissolving a conundrum.) Given that 'the present King of France' is a quantified noun phrase, (4) is ambiguous between:

(4a) There is exactly one present King of France and he is not bald.

(4b) It is not the case that there is exactly one present King of France and he is bald.

(4a) like (3), is false, because a necessary condition for its truth—i.e., that there is exactly one present King of France—is not satisfied. (4b), however, is true, and for precisely that reason, that there is not exactly one present King of France. The constituent existence clause is false, which renders true the negation of the whole. There is thus a sense in which (4) is true, and the threat to the law of excluded middle is skirted.

Descriptions can be meaningful without meaning any thing. Given semantic monism, this shows that descriptions do not function as referring expressions. Rather, they are quantified noun phrases. Thus, on Russell's view, there are two ways in which a thought or utterance can be about something: [a] a designator can refer to it directly (e.g., '*Two* is a positive integer'), and [b] a designator can denote it, in which case the relation between expression and denotation is mediated by the meanings of its constituent bits (e.g., '*The even prime* is a positive integer'). The following passage nicely conveys some of the semantic differences between denoting and referring:

> *Scott* is merely a noise or shape conventionally used to designate a certain person; it gives us no information about that person, and has nothing that can be called meaning as opposed to denotation.... But *the author of Waverley* is

not merely conventionally a name for Scott; the element of mere convention belongs here to the separate words *the* and *author* and *of* and *Waverley*. Given what these words stand for, *the author of Waverley* is no longer arbitrary.... A man's name is what he is called, but however much Scott had been called the author of Waverley, that would not have made him be the author; it was necessary for him actually to write Waverley, which was a fact having nothing to do with names. ([12], 258).

(Parenthetical remark omitted: "I neglect the fact, considered above, that even proper names, as a rule, really stand for descriptions." This gets us into an important issue in understanding Russell's thought during this period, but which is too tangential to our concerns to deal with in sufficient depth here. Epistemological considerations infect Russell's semantics, in that questions about reference are all tied up with questions about knowledge. He holds that one can only entertain object-dependent information about something one can be certain exists; and, strictly speaking, one cannot be certain of the existence of very much at all. Russell is propelled toward the claim that one can only refer to things that are immediately and currently present to mind. Eventually, this leads to Russell claiming that only 'this,' when used to refer to concurrent sense data, counts as a referring expression, and everything else really stands for a description. Today, it is generally acknowledged that Russell is running distinct kinds of questions together in this line of thought. The view of contemporary Russellians is essentially that captured by the above excerpt, according to which ordinary proper names do refer.)

Another line of support Russell offers for his theory consists of arguments against Meinong and Frege. Meinong has no good answer to the first puzzle above, about George IV's wondering whether Scott is the author of *Waverley*. His is a referential semantics, and informative identities are a problem for that view. Regarding the present King of France, a Meinongian answer is that since this is a potential subject of discourse, it subsists without being real. Russell alleges that Meinong breaches the law of excluded middle, that Meinong holds that the present King of France both exists and does not exist, that the round square both is and is not round. ([11]: 238) Quine (1953) pursues a slightly different line of attack premised on the question: just how many possible fat men are there in this doorway? The point here is that Meinong has no principled way to count non-actual objects, to distinguish one from another, and hence it is a disservice to thought to call such things 'objects.'

It is not clear that either accusation is fair, that Meinong's non-actual objects in fact have either of these defects. Even if the criticisms do apply to Meinong's theory, it is not obvious that they apply to any view that posits non-actual possibilities, and succeeding metaphysicians have developed more comprehensive theories of non-actual objects. However, it is clear that Meinong's naïve referential subject-predicate semantics is to be avoided. Again, metaphysics aside, Meinong is in need of a lesson in philosophical logic.

Russell gives two arguments against Frege ([11]: 238-40, 241-42). The first concerns non-denoting descriptions such as 'the present King of France;' the second is much more ambitious: built around the case of the description 'the first line of Gray's Elegy,' it is intended to demonstrate the untenablility of semantic dualism. The arguments are an inextricable tangle, and are, at best, inconclusive. Russell almost certainly confuses his first theory of denoting with Frege's sense-reference theory. The arguments produced here do seem to be those that caused him to reject his earlier theory, but they fail to engage with Frege's views, because of the unflinching assumption of semantic monism. It is assumed throughout that a referring expression that does not designate anything is nonsense, and that co-extensive referring expressions must make exactly the same contribution to propositional content. This is un-Fregean, on whose view it is coherent to wonder of a referring expression with a sense whether it, in fact, has a reference; and whose theory is premised on the claim that there can be an objective semantic difference between co-referential referring expressions. The presence of these un-Fregean assumptions strongly suggests that Russell fails to engage with Frege's views here.

To some extent, there is a straw target at work here. Even if Russell's arguments are valid, there are serious questions here about whether they refute anything Frege holds, and there are even more serious questions as to whether they refute any and all semantic dualist views. All that is addressed here is one particular approach to dualism. So the case is far from closed by Russell's arguments. The only clear upshot is that the sense-reference distinction is incompatible with semantic monism. Strictly speaking, there is no evidence that Russell understood Frege's position on senses, let alone offered a cogent argument against the position. The debate between the semantic monists and semantic dualists endures today.

Regardless, Russell's theory of descriptions is an important advance in the study of thought and language. Descriptions are insightfully and productively treated as quantified noun phrases. There are these two different semantic connections between thoughts and utterances and what they are about. Denoting expressions designate something in virtue of the meanings of their parts and the way in which they are structured; referring expressions refer to an object. It is important to attend to the differences between these two kinds of semantic connection. So, irrespective of whether Russell truly succeeds in accommodating Frege's data without embracing Frege's commitments, this is a valuable contribution to philosophy, a step forward in our understanding of reference and intentionality.

Within the analytic tradition, Russell ([11]) is the very paradigm of what philosophers ought to do. Semantic insight and conceptual clarity are earned by good solid philosophical logic. This semantic difference may look trivial on the surface, but failing to heed it could lead to erroneous inferences and to all manner of philosophical error.

As Russell mentions in the first and last pages of [11], the consequence of this distinction that excited him most is knowledge by description. Russell believed the discovery of denoting to afford the first satisfactory account of how it is that empirical knowledge can outstrip the bounds of perceptual experience, of how one can know things about contingent matters one has never encountered. In this sense, Russell's knowledge by description is akin to Kant's synthetic *a priori*: it is a genuinely third sort of knowledge, in addition to the traditional empiricists' knowledge by *a posteriori* causal connection between knower and known and the traditional rationalists' knowledge by *a priori* reasoning. This distinction is explored in depth in [12].

To sum up, we have just worked through two important differences between Frege and Russell: the first concerning the differences between monism and dualism, the second concerning the logical form of denoting expressions. Here comes a third: what to do about Russell's paradox. What, precisely, is logic?

V.iii *Philosophy of Logic*

Philosophy of logic deals with philosophical questions about the discipline of logic. Most generally, what *is* logic, this most formal and general of all disciplines? What is logic the study of, and by what method should it proceed? Precisely what does it mean to call something a *logical* truth, and what sorts of things are aptly so-called? The need for a philosophy of logic did not become explicit until the developments catalogued in this section. By this I do not mean to deny that there were substantive and important in-house disagreements within traditional logic. Intensionalists and extensionalists disagreed over whether inferences directly concern concepts or sets, for instance, and realists about logical truth disagreed with psychologistic theorists about the relation between human thought and the laws of logic. However, these are meta-disputes, in a way that certain debates in the early years of the twentieth century are not: the rival factions in these traditional disputes could agree entirely about the extent of logical truths, in a way that the opposing factions could not, in their response to the developments to be described here.

What I mean in saying that there is no philosophy of logic prior to the twentieth century is perhaps more clearly put as follows: from Aristotle to Frege, the only traditional answer to the central questions in the philosophy of logic is a naïve uncritical realism. Naïve uncritical realism is the view that logic is an *a priori* science whose truths are all self-evident, and to whose subject matter we rational agents all have access. The logical truths are those that we intuit to be self-evident. Logic studies the laws of truth, according to Boole, the laws on which all knowledge rests, according to Frege. The claim here is that one must reason in accord with such principles, or else one is prone to become ensnared in contradiction and incoherence; that is unobjectionable, but not yet anywhere near an acceptable philosophy of logic. It affords no means to settle disputes about controversial candidates for the status of logical truths, and some controversial candidates became crucial to the logicist project. In section III.i we worked through the way certain developments in the nineteenth century caused people to wonder about the foundations of mathematics. This section details a related story in which certain developments led people to wonder about the foundations of logic. Debates surrounding the proper understanding of these developments precipitated debates about what ought to count as logic, and thus made explicit the issues and questions subsequently known as the philosophy of logic.

Frege, by 1879, and Russell, some 20 years later, set out to formalize the study of logic, to develop new means for the study of logic, toward the end of reducing mathematics to logic. In the course of these projects, there was alot of thought given to what mathematics is, and to how to translate mathematics into some other format or discipline, but the fact is that not much thought was given as to what, precisely, the discipline of logic is. The founders of modern logic, too, were naïve uncritical realists about logic. Many years after it was recognized that, as Russell puts it, "self-evidence is often a mere will-o'-the-wisp which is sure to lead us astray if we take it as our guide" ([10]: 223) in mathematics, it was still assumed that self-evidence could play that role in grounding the truths of logic. However, naïve uncritical realism was rendered obsolete by two developments, in the first five years of the twentieth century: Russell's paradox and Zermelo's axiom of choice.

The notion of a set is central to traditional logic, and is crucial to the logicist project. As conceived by both Frege and Russell, logicism helps itself to the notion of a set, defines 'number' in terms of equinumerous sets, and goes on to demonstrate that all arithmetic operations and relations can be translated into, or reduced to, logical operations performed on these kinds of set. Given a neat relation between sets and predicates, properties of and relations between numbers can be formalized and studied within the predicate calculus, within modern logic. Sets have been around logic more or less from the very beginning, providing a clear extensional means of talking about more obscure concepts. As a self-evident point of logic, it was assumed that there is a simple one-one relation between intensions and extensions—for every coherent intension, there corresponds a set of things in the extension of the term, and for everything that deserves to be called a set, there corresponds a clear intension.

Kablooie! Russell's paradox and Zermelo's axiom each undermined one arm of this bi-directional intension-extension, property-set relation. Zermelo's axiom of choice asserts that, for every set of sets, there is a set containing one member of each of those sets. This asserts the existence of a set for which there is no corresponding intension, no property true of all and only its members. Zermelo demonstrated that this axiom had been relied on, unacknowledged, in many important proofs. Russell and Whitehead came to see that this axiom is required in order to reduce mathematics to logic. However, there were dissenters, right from the start, who thought that it is a mistake to call such an assertion, in no way self-evident, an 'axiom.' Tradition demanded a property in order for there to be a set; Zermelo's axiom flies in the face of that demand. Peano, for instance, expressed skepticism as to whether the axiom of choice is true, *a fortiori* whether it is a truth of logic. In response, Zermelo replied with the hardest question one could ask at the time: "How does Peano arrive at his [axioms], and how does he justify [them] ...?" (Coffa, 1991: 118) Although there is nothing obviously self-evident about the axiom of choice, it is enormously useful in derivations, and essential to the project of establishing that mathematics is a branch of logic. Is it a part of logic, then? Is it part of what we mean by 'set'? What is a set? What is logic?

Russell's paradox is the opposite of the axiom of choice in that it involves a coherent intension or property to which no extension, no set, could possibly correspond. Russell's paradox is best illustrated via an example. Call the sets that are not members of themselves 'normal,' and sets that are members of themselves 'non-normal.' Thus, R, the set of all rocks, is normal, because it is not itself a rock, whereas N, the set of all things that are not rocks, is non-normal, because it is itself not a rock, and hence is itself among its elements. Now call the set of all normal sets, of sets that do not themselves satisfy the intension that defines their membership, 'E.' So R is a member of E, but N is not. If we then ask whether E belongs to itself, whether E is normal or non-normal, this lands us in a big problem. Each alternative leads to its opposite. If it is a member of itself then it does not possess its defining property, whereas if it is not a member of itself then it does possess that property. Thus the paradox: E is normal if and only if it is non-normal. (Here is a simpler example: consider a barber who shaves all and only those in a certain town who do not shave themselves. If we then ask whether this barber shaves himself, we have a paradox: if he does shave himself, then he is not one of those he shaves, and if he does not, then he is.)

Russell's paradox created an air of despair in the logicists' camp, from which Frege never recovered. The traditional notion of a set, of the one-one correlation between the intension and extension of a coherent predicate, is central to the logicist project, and this paradox yanked out the rug from under it. "Without a single object to represent an extension," says Russell (1903: 515), "mathematics crumbles." Frege admitted that logicism had been refuted by Russell's paradox: mathematics can be reduced to logic only if 'set' is a logical notion; Russell has inadvertently proven that it is not. Frege sank into despair, and gave up on his lifelong project.

Russell's response is quite different. He develops a theory whose aim is to rule out as unintelligible the offending sets. This is Russell's theory of types, and it is a milestone in the development of set theory. The problem with the paradox, as with a number of others Russell (1908: 59-61) outlines, is that the intension or property seems to be well-formed, but is not. There are different types of sets, and different types of members, and a certain type of set can only have one specific type of element as members. A first-order set can only have objects as members, a second-order set can only have sets of objects as members, a third-order set can only have sets of sets of objects as members, and so on up. (For instance, the Liberal party is an example of a first-order set, containing as members a bunch of individuals; the set of Canadian political parties is a second-order set, containing as members sets of individuals; the set containing as elements the sets of sets that are the Canadian political parties, the British political parties, the French political parties, and so on, is a third-order set; and there is no limit to how high the hierarchy can go.) So there are different types of element, and different types of set, and tight restrictions on what type of elements a set can have as members.

The paradoxes result from transgressing this stricture, from the assumption that a set could have both individuals and sets as members, so that it could have itself as a member. (In the analogy, there just simply cannot be a barber who shaves all and only those who do not shave themselves, any more than there could be a round square.) The assertion that a set is a member of itself is not true or false but ruled out as meaningless; the offending intensions or properties are ruled out as incoherent.

Now this raises questions very similar to those raised by the axiom of choice. Is all this really part of what we mean by 'set'? Is the theory of types self-evident? Is this kind of set theory a branch of logic? Again, it is required to save the logicist thesis; but the logicist thesis is clearly being redefined here, plank by plank. Adding the axiom of choice and the theory of types to logic is not outlandish, not obviously or illicitly *ad hoc*; but it is marching away from the original spirit of the project, which is about demonstrating that the truths of mathematics rest on the firmest foundations possible. At the very least, logicism has been redefined by Russell. In general, the consensus is that logicism is effectively refuted by Godel's (1931) famous incompleteness results. However, Russell remained unrepentant (1959: 78), continuing to hold that the theory of types effectively protects logicism from Godel's results.

The claim that there is no explicit philosophy of logic throughout the tradition is precisely the claim that there is nothing within the canon for theorists to appeal to, in order to justify their position on these questions. There are only rival factions with distinct but equally vague and unsystematic intuitions about what is self-evident, about what counts as logic. Frege's position is that these gerrymandered set-theories are clearly and obviously not logic. His philosophy of logic is a series of consistent but vague gestures toward the realist party line. He does have deep and insightful criticisms of alternative views, of the psychologistic tendencies in the views of his idealist and empiricist contemporaries, but he has little of substance in the way of positive views. (I suspect that this is something of a characteristic point about philosophy of logic here: the strongest points in favor of any view are negative, concerning problems into which the alternatives lead. Perhaps this is a feature in common among the deepest of the questions of philosophy. I say some things to substantiate the former conjecture, but not the latter, in the next few paragraphs.)

So, clearly, Frege and Russell have different conceptions of logic. Coffa (1991: 118) speculates that the logicist thesis is a regulative maxim, a guide to characterizing logic, in Russell's work; i.e., whereas Frege has a conception of what it would take to reduce mathematics to logic from the outset, which is precise enough to have been refuted by Russell's paradox, Russell's logicist thesis is stipulative, in that 'logic' for Russell just means 'whatever is needed to shore up the foundations of mathematics.' That conjecture is, to say the least, contentious. However, it is clear that Russell is a Platonic dialogue unto himself (as Whitehead once famously remarked) on some central questions in the philosophy of logic. The participants in this particular dialogue are Russell the realist and Russell the empiricist. Epistemological considerations push him toward realism; realism is one of the things which propelled him toward his rejection of idealism, and is defended eloquently in his (1912). In his ([13]: 265, 276) he explains how recent developments in mathematics and logic have refuted empiricism and proven the need for a non-empirical source of knowledge about non-empirical objects. However, Russell the empiricist is swayed by at least two kinds of worries: the metaphysical question of exactly what, and where, these non-empirical objects are, and the epistemological question of precisely how we can know anything about them. There are some strong *a priori* reasons pushing Russell toward realism about logic, but pushing against them is the empiricists' conviction that *a posteriori* evidence ought to be the ultimate metaphysical arbiter.

The following is an early passage that attests to the role of empirical considerations in the *a priori* sciences for Russell:

> There is a greatest of all infinite numbers, which is the number of all things together, of every sort and kind. It is obvious that there cannot be a greater number than this, because, if everything has been taken, there is nothing left to add. ([10]: 229)

Throughout his career, and despite his realism, logic and mathematics are, in part, empirical studies for Russell, in a way that is antithetical to the classical sort of realism that Frege espouses. Logic is the same kind of science as zoology ([14]: 280), only its subject matter is more abstract and general. The reasons for accepting something as a logical truth, for Russell, are, partly, inductive: something can earn its way as an axiom not in virtue of its content but in virtue of its effects, its place in a productive system. The truths of logic themselves are akin to empirical generalizations: "...the law of contradiction is ... a fact concerning the things in the world." (1912: 89) There is no difference in kind between the meaning of an *a priori* truth and an empirical generalization (1912: 106). Logic is the study of abstract forms, which play some crucial constitutive role in *a priori* knowledge, but these forms somehow have to be chunks of this world. How else could we have knowledge of them?

With much of this, the realist Frege would disagree. Empirical generalizations are contingent, but the law of contradiction is fashioned from harder metal than that. Realists hold that the truths of logic are mind-independent, that they are indifferent to empirical and psychological contingencies. The facts could change, and humans could come to think differently; but neither of these considerations is even relevant to whether something is a logical truth, for a realist.

These tensions in Russell's view betray the difficulty of some of these questions in the philosophy of logic. Thinking about the indifference of logical truths to human psychology and to contingent facts propels one toward realism, toward the claim that the truths of logic are thoroughly mind-independent, but it is very hard to explain precisely the content of this realism. It seems to rest on a mysterious kind of vision of a mysterious sort of creature, and there are good sound reasons to avoid positing such shadowy, unexplained objects and faculties of mind. Other philosophies of logic have been developed in the twentieth century, of which the most prevalent are varieties of conventionalism and naturalism. Conventionalism, inspired by Wittgenstein and developed by the positivists, is built around the suggestion that it is our conventions to use terms in the way that we do that determines the logical truths. On this approach, the laws of logic are true by agreement, akin to 'Drive on the right side of the road around here.' Naturalism is the view that the laws of logic are descriptive regularities, empirical generalizations, like F=MA. There are different varieties of naturalism about logic, ranging from Kantian-psychologistic interpretations, according to which logical truths are determined by the way we humans reckon, to a more Russellian sort of naturalism, which explains the logical truths by appeal to how things are in the world. Perhaps Darwin provides a bridge between these two naturalisms: the empirical contingencies which have determined how we have evolved also and thereby determine how we think and ought to infer. On this intermediate naturalism, the truths of logic look a lot like biological laws.

In general, the strength of these non-realist philosophies of logic lies in avoiding the difficult metaphysical and epistemological problems that confront the realist, and their weakness lies in their inability to explain the empirical and psychological transcendence of the truths of logic. These views do not face the most difficult questions for realism; but it is far from clear that this is because they address the issues from which those questions stem, as opposed to just avoiding the difficult questions. Many varieties of views in the philosophy of logic have been developed and defended along these dimensions. Debate continues as to which sort of view provides the best way to think about logic.

VI. The Legacy of Logicism

This concludes the story of how mathematical logic became such a central instrument in the analysis of language; and among the greatest legacies of logicism is the very fact that the logic the logicists developed has become such a prevalent and productive instrument. I will end with a brief statement of the legacy of Frege and Russell, of the impact that their creation and pursuit of this style of philosophical analysis has had on subsequent philosophy. It is grouped under three headings: form, content, and method.

[1] Form
Few, if any, can parallel their contributions to the formal study of inference. Many of their contemporaries, in mathematics and in philosophy, were pursuing studies that would lead to important parts of modern logic; but, for the reasons discussed herein, Frege has the strongest claim to be the founder of modern logic, and Russell did more than anyone else to further its ends and to expound on its virtues. Frege and Russell pioneer the application of mathematical tools and techniques to the study of information, and that has been one of the most potent philosophical and technological ideas of recent centuries.

[2] Content
As for content, this brings us from logic to philosophy. The depth of their semantic analyses, first in mathematics and then more broadly, is unprecedented. Frege and Russell are the first philosophers of language, and, on many ways of measuring greatness, still the best. They identified, articulated, and made progress toward, fundamental questions about truth, meaning, reference, and intentionality. Their mark is still being unpacked, in many corners of the philosophical study of language and mind.

[3] Method
Those first two aspects compose the content of their contribution to philosophy; this last aspect concerns the way in which that contribution is made. This aspect of their contribution is both the least substantive and the most pervasive. Frege and Russell defined the analytic method in philosophy, a distinctive approach to philosophical questions, which has affected the way in which issues are addressed in all corners of the discipline. Language and logic aside, some of the best work in the twentieth century, in more or less every sub-discipline within academic philosophy, has been influenced by the way in which the logicists do philosophy. Though logicism dies off, much of the work done toward that end lives on.

Works Cited

Arnauld, A., and Nicole, P. 1662. *The Art of Thinking: Port Royal Logic*. Indianapolis: Bobbs-Merril, 1964.
Ayer, A.J. 1936. *Language, Truth, and Logic*. New York: Dover.
Boole, G. 1854. *The Laws of Thought*. Chicago: Open Court, 1916.
Brentano, F. 1884. *Psychology from an Empirical Standpoint*. London: Routledge, 1973.
Coffa, J.A. 1991. *The Semantic Tradition from Kant to Carnap*. Cambridge: Cambridge University Press.
Frege, G. 1884. *The Foundations of Arithmetic*. Translated by J.L. Austin. Oxford: Blackwell, 1953.
—. 1903. *The Basic Laws of Arithmetic*, 2nd volume. Translated by M. Furth. Los Angeles: University of California Press, 1964.
—. 1979. *Posthumous Writings*. Translated by P. Long and P. White. Oxford: Blackwell.
—. 1980. *Philosophical and Mathematical Correspondence*. Translated by H. Kaal. Oxford: Blackwell.
Godel, K. 1931. "Some Metamathematical Results on Completeness and Consistency." Translated by J. van Heijenoort, in *From Frege to Godel: A Sourcebook in Mathematical Logic*. Cambridge: Harvard University Press, 1967.
Kant, I. 1784. *Critique of Pure Reason*. Translated by Norman Kemp Smith. London: MacMillan, 1929.
Kenny, A. 1995. *Frege*. London: Penguin.
Kripke, S. 1972. *Naming and Necessity*. Cambridge: Harvard University Press.
Leibniz, G. 1966. *Logical Papers*. Translated by G.H.R. Parkinson. Oxford: Clarendon.
Locke, J. 1690. *An Essay Concerning Human Understanding*. London: Dent, 1947.
Meinong, A. 1904. "The Theory of Objects," in *Realism and the Background of Phenomenology*. Edited by R.M. Chisholm. Glencoe: The Free Press, 1960.
Mill, J.S. 1830. *A System of Logic*. London: Longmans, 1947.
Moore, G.E. 1903. "The Refutation of Idealism." *Mind* vol. XII.
Putnam, H. 1975. "The Meaning of 'Meaning,'" in *Language, Mind, and Knowledge*. Edited by K. Gunderson. Minneapolis: University of Minnesota Press.

Quine, W.V. 1953. "On What There Is," in *From a Logical Point of View*. Cambridge: Harvard University Press.
—. 1956. "Quantifiers and Propositional Attitudes." *Journal of Philosophy* vol. 53.
—. 1960. *Word and Object*. Cambridge: MIT Press.
Russell, B. 1902. "The Study of Mathematics," in *Mysticism and Logic and other essays*. London: Allen and Unwin, 1917.
—. 1903. *The Principles of Mathematics*. London: Allen and Unwin.
—. 1908. "Mathematical Logic as Based on the Theory of Types," in *Logic and Knowledge*. Edited by R.C. Marsh. London: Unwin Hyman, 1956.
—. 1912. *The Problems of Philosophy*. London: Allen and Unwin.
—. 1918. "The Philosophy of Logical Atomism." in *Logic and Knowledge*. Edited by R.C. Marsh. London: Unwin Hyman, 1956.
—. 1959. *My Philosophical Development*. London: Unwin Hyman.
—. 1967-9. *The Autobiography of Bertrand Russell*. London: Allen and Unwin.
Russell, B., and Whitehead, A.N. 1910. *Principia Mathematica*. Cambridge: Cambridge University Press.
Urmson, J.O. 1956. *Philosophical Analysis: Its Development between the Two World Wars*. Oxford: Clarendon.
Wittgenstein, L. 1921. *Tractatus Logico-Philosophicus*. London: Routledge.
—. 1953. *Philosophical Investigations*. Oxford: Blackwell.
—. 1961. *Notebooks 1914-1916*. New York: Harper and Row.

WORKS OF
GOTTLOB FREGE

I
CONCEPTUAL NOTATION

A Formula Language of Pure Thought Modelled upon the Formula Language of Arithmetic

Preface

The apprehension of a scientific truth proceeds, as a rule, through several stages of certainty. First guessed, perhaps, from an inadequate number of particular cases, a universal proposition becomes little by little more firmly established by obtaining through chains of reasoning, a connection with other truths — whether conclusions which find confirmation in some other way are derived from it; or, conversely, whether it comes to be seen as a conclusion from already established propositions. Thus, on the one hand, we can ask by what path a proposition has been gradually established; or, on the other hand, in what way it is finally most firmly establishable. Perhaps the former question must be answered differently for different people. The latter [question] is more definite, and its answer is connected with the inner nature of the proposition under consideration.

The firmest method of proof is obviously the purely logical one, which, disregarding the particular characteristics of things is based solely upon the laws on which all knowledge rests. Accordingly, we divide all truths which require a proof into two kinds: the proof of the first kind can proceed purely logically, while that of the second kind must be supported by empirical facts. It is quite possible, however, for a proposition to be of the first sort and still be one that could never come to the consciousness of a human mind without activity of the senses.* Therefore, not the psychological mode of origin, but the most perfect method of proof underlies the classification.

Now, while considering the question to which of these two kinds [of truths] do judgements of arithmetic belong, I had first to test how far one could get in arithmetic by means of logical deductions alone, supported only by the laws of thought, which transcend all particulars. The procedure in this effort was this: I sought first to reduce the concept of ordering-in-a-sequence to the notion of *logical* ordering, in order to advance from here to the concept of number. So that something intuitive could not squeeze in unnoticed here, it was most important to keep the chain of reasoning free of gaps. As I endeavoured to fulfil this requirement most rigorously, (I

---

* Since without sense perception no mental development is possible for beings known to us, the latter point holds for all judgements.

found an obstacle in the inadequacy of the language;) despite all the unwieldiness of the expressions, the more complex the relations became, the less precision — which my purpose required — could be obtained. From this deficiency arose the idea of the "conceptual notation" presented here. Thus, its chief purpose should be to test in the most reliable manner the validity of a chain of reasoning and expose each presupposition which tends to creep in unnoticed, so that its source can be investigated. For this reason, I have omitted the expression of everything which is without importance for the chain of inference. In §3, I have designated by *conceptual content* that which is of sole importance for me. Hence, this must always be kept in mind if one wishes to grasp correctly the nature of my formula language. Also, the name "Conceptual Notation" resulted from this. Since I limited myself, for the present, to the expression of relations which are independent of the particular state of things, I was also able to use the expression "formula language of pure thought". The modelling upon the formula language of arithmetic to which I have alluded in the title refers more to the fundamental ideas than to the detailed structure. The farthest thing from my mind was any effort to establish an artificial similarity through the interpretation of a concept as the sum of its characteristic marks. The most immediate point of contact between my formula language and that of arithmetic is in the way letters are used.

I believe I can make the relation of my "conceptual notation" to ordinary language clearest if I compare it to the relation of the microscope to the eye. The latter, because of the range of its applicability and because of the ease with which it can adapt itself to the most varied circumstances, has a great superiority over the microscope. Of course, viewed as an optical instrument it reveals many imperfections, which usually remain unnoticed only because of its intimate connection with mental life. But as soon as scientific purposes place strong requirements upon sharpness of resolution, the eye proves to be inadequate. On the other hand, the microscope is perfectly suited for just such purposes; but, for this very reason, it is useless for all others.

Similarly, this "conceptual notation" is devised for particular scientific purposes; and therefore one may not condemn it because it is useless for other purposes. Even if it fulfils its purposes in some measure, one may still fail to find new truths in my work. I would nevertheless take comfort in the conviction that an improvement in method also advances science. Bacon, after all, held that it is more important to invent a means by which everything can be discovered easily than to discover some particular thing; and, in fact, all great scientific advances in recent times have had their origin in an improvement of method.

Leibniz also recognized – perhaps overestimated – the advantages of an adequate method of notation. His idea of a universal characteristic, a *calculus philosophicus* or *ratiocinator*,* was too ambitious for the effort to realize it to go

---

* On that point see Trendelenburg, *Historische Beiträge zur Philosophie*, Volume 3 [(1867) "Über Leibnizes Entwurf einer allgemeinen Charakteristik", pp.1–47.]

beyond the mere preparatory steps. The enthusiasm which overcomes its [would be] creator when he considers what an immense increase in the mental power of mankind would result from a method of notation which fits things themselves lets him underestimate the difficulty which such an undertaking confronts. But even if this high aim cannot be attained in one try, we still need not give up hope for a slow, stepwise approximation. If a problem in its complete generality appears unsolvable, we have to limit it provisionally; then, perhaps, it will be mastered with a gradual advance. We can view the symbols of arithmetic, geometry, and chemistry as realizations of the Leibnizian idea in particular areas. The "conceptual notation" offered here adds a new domain to these; indeed, the one situated in the middle adjoining all others. Thus, from this starting point, with the greatest expectation of success, we can begin to fill in the gaps in the existing formula languages, connect their hitherto separate domains to the province of a single formula language and extend it to fields which up to now have lacked such a language.

I am sure that my "conceptual notation" can be successfully applied wherever a special value must be placed upon the validity of proofs, as in laying the foundation of the differential and integral calculus.

It appears to me to be still easier to extend the area of application of this formula language to geometry. We should only have to add a few symbols for the intuitive relations that occur there. In this way, we should acquire a kind of *analysis situs*.

The transition to pure kinematics and further to mechanics and physics might follow here. In the latter fields, where besides necessity of thought, physical necessity asserts itself, a further development of the mode of notation with the advancement of knowledge is easiest to foresee. But this is no reason to wait until such transformations appear to have become impossible.

If it is a task of philosophy to break the power of the word over the human mind, uncovering illusions which through the use of language often almost unavoidably arise concerning the relations of concepts, freeing thought from that which only the nature of the linguistic means of expression attaches to it, then my "conceptual notation", further developed for these purposes, can become a useful tool for philosophers. Certainly, it also does not reproduce ideas in pure form either, and this is probably inevitable for a means of thought expression outside of the mind; but on the one hand, we can limit these discrepancies to the unavoidable and harmless; and on the other hand, merely because they are of a completely different kind from those [discrepancies] peculiar to [ordinary] language, they provide a protection against onesided influence of one such means of expression.

The mere invention of this "conceptual notation", it seems to me, has advanced logic. I hope that logicians, if they do not allow themselves to be frightened off by the first impression of unfamiliarity, will not refuse their assent to the innovations to which I have been driven by a necessity inherent in the subject matter itself. These deviations from the traditional find their justification in the fact that logic up to now has always confined itself too closely to language and grammar. In particular, I believe that the replacement of the concepts of *subject* and *predicate* by *argument* and *function* will prove itself in the long run. It is easy to see how regarding a content as a function of an argument leads to the formation of concepts. Furthermore, the demonstration of the connection between the meanings of the words: if, and, not, or, there exists, some, all, and so forth, may deserve notice.

Only the following further things require mention here:

The restriction, declared in §6, to a single mode of inference is justified by the fact that in laying the *foundation* of such a conceptual notation the primitive components must be chosen as simple as possible if perspicuity and order are to be created. This does not preclude the possibility that, *later*, transitions from several judgements to a new one, which are possible by this single mode of inference in only an indirect way, be converted into direct ways for the sake of abbreviation. Indeed, this may be advisable in a later application. In this way, then, further modes of inference would arise.

I noticed only later that formulas (31) and (41) can be combined into the single formula

$$\vdash (\mathop{\rlap{\,\rule[0.5ex]{0.8em}{0.4pt}}\rule[1ex]{0.8em}{0.4pt}} a \equiv a)$$

which makes even more simplifications possible.*

Arithmetic, as I said at the beginning, was the starting point of the train of thought which led me to my "conceptual notation". I intend, therefore, to apply it to this science first, trying to analyse its concepts further and provide a deeper foundation for its theorems. For the present, I have presented in the third chapter some things which move in that direction. Further pursuit of the suggested course – the elucidation of the concepts of number, magnitude, and so forth – is to be ths subject of further investigations which I shall produce immediately after this book.

Jena, 18 December 1878

---

* Editor's note: This refers to a subsequent chapter of this work, not reprinted here.

I. Definition of the Symbols

§1. The symbols customarily used in the general theory of magnitude fall into two kinds. The first consists of the letters, each of which represents either a number left undetermined or a function left undetermined. This indeterminateness makes it possible to use letters for the expression of the general validity of propositions, as in

$$(a + b)c = ac + bc.$$

The other kind consists of such symbols as $+$, $-$, $\sqrt{}$, 0, 1, 2; each of which has its own specific meaning.

I adopt this fundamental idea of distinguishing two kinds of symbols, which unfortunately is not strictly carried through in the theory of magnitudes,* in order to use it for the more inclusive domain of pure thought in general. I therefore divide all the symbols I employ into those which one can take to signify various things and those which have a completely fixed sense. The first are the letters, and these are to serve mainly for the expression of generality. But we must insist that a letter, for all its indeterminateness, should retain throughout the same context the meaning which we first gave it [in that context].

The judgement

§2. A judgement will always be expressed with the aid of the symbol

which stands to the left of the symbol or combination of symbols giving the content of the judgement. If we omit the small vertical stroke at the left end of the horizontal one, then the judgement is to be transformed into a *mere combination of ideas* of which the writer does not state whether or not he acknowledges its truth. For example, let

⊢——— A

mean the judgement: "Opposite magnetic poles attract each other."†
Then

——— A

will not express this judgement, but should simply evoke in the reader the idea of the reciprocal attraction of opposite magnetic poles, perhaps, say, in order to derive

---

* Consider 1, log, sin, lim.

† I use capital Greek letters [in this case capital *alpha*] as abbreviations to which the reader may attribute an appropriate sense if I do not specifically define them.

some conclusions from it and with these test correctness of the thought. We paraphrase in this case by means of the words *"the circumstance that"* or *"the proposition that"*.

Not every content can become a judgement by placing ⊢——— before its symbol; for example, the idea "house" cannot. We therefore distinguish *assertible* and *unassertible* contents.*

The horizontal stroke, which is part of the symbol ⊢——— *, ties the symbols which follow it into a whole; and the assertion, which is expressed by means of the vertical stroke at the left end of the horizontal one, relates to this whole.* Let us call the horizontal stroke the *content stroke*, the vertical one the *judgement stroke*. The content stroke serves also to relate any sign to the whole formed by the symbols that follow the stroke. *Whatever follows the content stroke must always have an assertible content.*

§3. A distinction between *subject* and *predicate* does *not occur* in my way of representing a judgement. To justify this, I note that the contents of two judgements can differ in two ways: first, it may be the case that [all] the consequences which can be derived from the first judgement combined with certain others can always be derived also from the second judgement combined with the same others; secondly, this may not be the case. The two propositions, "At Plataea the Greeks defeated the Persians." and "At Plataea the Persians were defeated by the Greeks.", differ in the first way. Even if one can perceive a slight difference in sense, the agreement [of sense] still predominates. Now I call the part of the content which is the *same* in both the *conceptual content*. Since *only this* is meaningful for [our] "conceptual notation", we need not distinguish between propositions which have the same conceptual content. If one says, "The subject is the concept with which the judgement is concerned.", this applies also to the object. Therefore, we can say only: "The subject is the concept with which the judgement is chiefly concerned." In [ordinary] language, the subject-place has the significance in the word-order of a *special* place where one puts what he wishes the listener to particularly heed. (See also §9.) This can serve, for instance, to indicate a relation of this judgement to others, thus facilitating for the listener an understanding of the whole context. Now all aspects of [ordinary] language which result only from the interaction of speaker and listener – for example, when the speaker considers the listener's expectations and tries to put them on the right track even before speaking a [complete] sentence – have nothing corresponding to them in my formula language, because here the only thing considered in a judgement is that which influences its

---

* On the other hand, the circumstance that there are houses (or there is a house) would be an assertible content. (See §12.) But the idea "house" is only a part of this. In the proposition, "Priam's house was made of wood.", we could not put "the circumstance that there is a house" in place of "house".

possible consequences. Everything necessary for a correct inference is fully expressed; but what is not necessary usually is not indicated; *nothing is left to guessing.* In this I strictly follow the example of the formula language of mathematics, in which, also, one can distinguish subject and predicate only by doing violence [to the language]. We can imagine a language in which the proposition, "Archimedes perished at the conquest of Syracuse.", would be expressed in the following way: "The violent death of Archimedes at the conquest of Syracuse is a fact." Even here, if one wishes, he can distinguish between subject and predicate; but the subject contains the whole content, and the predicate serves only to present this as a judgement. *Such a language would have only a single predicate for all judgements; namely, "is a fact".* We see that here we cannot speak of subject and predicate in the usual sense. *Our "conceptual notation" is such a language, and the symbol* ⊢——— *is its common predicate for all judgements.*

In my first draft of a formula language, I was misled by the example of [ordinary] language into forming judgements by combining subject and predicate. I soon became convinced, however, that this was an obstacle to my special goal and led only to useless prolixity.

§4. The following remarks are intended to explain the significance, for our purposes, of the distinctions which people make with regard to judgements.

People distinguish *universal* and *particular* judgements: this is really not a distinction between judgements, but between contents. *They should say, "a judgement with a universal content", "a judgement with a particular content".* These properties belong to the content even when it is put forth, not as a judgement, but as an [unasserted] proposition. (See §2.)

The same holds for negation. For example, in an indirect proof we say, "Suppose that the line segments *AB* and *CD* were not equal." Here the content – that line segments *AB* and *CD* are not equal – contains a negation; but this content, although it could be a judgement, is not presented as a judgement. Negation attaches therefore to the content, whether or not it occurs as a judgement. I therefore consider it more appropriate to regard negation as a characteristic of an *assertible content.*

The distinction of categorical, hypothetical, and disjunctive judgements appears to me to have only a grammatical significance.*

The apodictic judgement is distinguished from the assertoric in that the apodictic suggests the existence of general judgements from which the proposition can be inferred, while the assertoric lacks such an indication. If I call a proposition necessary, I thereby give a hint about my grounds for judgement. *But since this does not affect the conceptual content of the judgement, the apodictic form of judgement has for us no significance.*

---

* The reason for this will be brought out by the whole of this work.

If a proposition is presented as possible, the speaker is either refraining from judgement and indicating that he knows no laws from which the negation [of the proposition] would follow; or else he is saying that the universal negation of the proposition is false. In the latter case, we have what is usually called a *particular affirmative judgement*. (See §12.) "It is possible that the earth will someday collide with another heavenly body." is an example of the first case; and "A cold can result in death." is an [example] of the second case.

Conditionality

§5. If A and B stand for assertible contents (§2), there are the following four possibilities:

(1) A is affirmed and B is affirmed
(2) A is affirmed and B is denied
(3) A is denied and B is affirmed
(4) A is denied and B is denied.

Now,

stands for the judgement that *the third of these possibilities does not occur, but one of the other three does*. Thus, the denial of

signifies that the third possibility occurs, that therefore A is denied and B is affirmed.

From among the cases in which

is affirmed, we stress the following:

(1) A must be affirmed. Then the content of B is entirely immaterial. For example, let ├──── stand for $3 \times 7 = 21$ and let B stand for the circumstance that the sun is shining. Here only the first two of the four cases mentioned are possible. There need not exist a causal connection between the two contents.

(2) B is to be denied. Then the content of A is immaterial. For example, let B stand for the circumstance that perpetual motion is possible, and A the circumstance that the universe is infinite. Here only the second and fourth of the four cases are possible. There need be no causal connection between A and B.

(3) We can make the judgement

Conceptual Notation 101

without knowing whether A and B are to be affirmed or denied. For example, let B stand for the circumstance that the moon is in quadrature [with the sun], and A the circumstance that it appears as a semicircle. In this case, we can translate

$$\vdash\!\!\begin{array}{c}\rule{1em}{0.4pt}\ A\\ \rule{1em}{0.4pt}\ B\end{array}$$

with the aid of the conjunction "if": "if the moon is in quadrature [with the sun], it appears as a semicircle." The causal connection implicit in the word "if", however, is not expressed by our symbols, although a judgement of this kind can be made only on the basis of such a connection; for this connection is something general, but at this point we do not yet have an expression for generality. (See §12.)

Let the vertical stroke which connects the two horizontal ones be called the *conditional stroke*. The part of the upper horizontal stroke situated to the left of the conditional stroke is the content stroke for the meaning, just explained, of the symbol combination

$$\begin{array}{c}\rule{1em}{0.4pt}\ A\\ \rule{1em}{0.4pt}\ B;\end{array}$$

to this is attached every symbol which is intended to relate to the content of the expression as a whole. The part of the horizontal stroke lying between A and the conditional stroke is the content stroke of A. The horizontal stroke to the left of B is the content stroke of B.

Accordingly, it is easy to see that

$$\vdash\!\!\begin{array}{c}\rule{1em}{0.4pt}\ A\\ \rule{1em}{0.4pt}\ B\\ \rule{1em}{0.4pt}\ \varGamma\end{array}$$

denies the case in which A is denied and B and $\varGamma$ are affirmed. We must think of this as constructed from

$$\begin{array}{c}\rule{1em}{0.4pt}\ A\\ \rule{1em}{0.4pt}\ B\end{array}$$

and $\varGamma$ in the same way as

$$\begin{array}{c}\rule{1em}{0.4pt}\ A\\ \rule{1em}{0.4pt}\ B\end{array}$$

is constructed from A and B. Therefore, we have first the denial of the case in which

$$\begin{array}{c}\rule{1em}{0.4pt}\ A\\ \rule{1em}{0.4pt}\ B\end{array}$$

is denied and $\varGamma$ is affirmed. But the denial of

$$\begin{array}{c}\rule{1em}{0.4pt}\ A\\ \rule{1em}{0.4pt}\ B\end{array}$$

means that A is denied and B is affirmed. From this is obtained what is given above. If a causal connection is present, we can also say, "A is the necessary consequence of B and Γ"; or, "If the circumstances B and Γ occur, then A occurs also."

We can see just as easily that

denies the case in which B is affirmed, but A and Γ are denied. If we assume a causal connection between A and B, we can translate [the formula]: "If A is the necessary consequence of B, we can infer that Γ occurs."

§6. A result of the definition [of the conditional stroke] given §5 is that from the two judgements

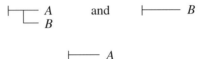

the new judgement $\vdash\!\!-\!\!- A$

follows. Of the four cases enumerated above, the third is excluded by

$\vdash\!\!\sqsubset\!\! \begin{array}{c} A \\ B \end{array}$

but the second and fourth [cases are excluded] by

$\vdash\!\!-\!\!- B$

so that only the first [case] remains.

We could write this inference perhaps this way:

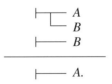

This would be laborious if long expressions stood in the places of A and B, because each of them would have to be written twice. I therefore use the following abbreviation: Each judgement which occurs in the context of a proof is labelled with a number which is placed to the right of the judgement at its first occurrence. Now, suppose for example that the judgement

– or one containing it as a special case – has been labelled with X. Then I write the inference this way:

$$(X): \dfrac{\vdash B}{\vdash A}$$

Here it is left to the reader to construct the judgement

$$\vdash \begin{array}{l} A \\ B \end{array}$$

from $\vdash B$ and $\vdash A$, and to see if it tallies with the cited judgement X.

If, for example, the judgement $\vdash B$ was labelled XX, then I also write the same inference this way:

$$(XX):: \dfrac{\vdash \begin{array}{l} A \\ B \end{array}}{\vdash A.}$$

Here the double colon indicates that $\vdash B$, which is cited only by means of XX, must be formed, by a method other than the one above, from the two judgements which are written out.

Further, if, say, the judgement $\vdash \Gamma$ were labelled with XXX, I would abbreviate the two inferences

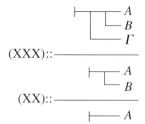

even more like this

$$(XX, XXX):: \dfrac{\vdash \begin{array}{l} A \\ B \\ \Gamma \end{array}}{\vdash A.}$$

In logic people enumerate, following Aristotle, a whole series of modes of inference. I use just this one – at least in all cases where a new judgement is derived from more than one single judgement. We can, of course, express the truth implicit in another mode of inference in one judgement of the form: if *M* holds and *N* holds, then *Λ* holds as well; in symbols:

From this judgement plus |—— *N* and |—— *M*, |—— *Λ* follows as above. In this way, an inference using any mode of inference can be reduced to our case. Accordingly, since it is possible to manage with a single mode of inference, perspicuity demands that we do so. Otherwise, there would be no reason to stop with the Aristotelian modes of inference; instead, we could go on adding new ones indefinitely: we could make a special mode of inference of each of the judgements expressed in formulas in §§13 to 22.* *This restriction to a single mode of inference, however, is in no way intended to express a psychological proposition, but merely to settle the question of the most expedient form* [*of our "conceptual notation"*]. Some of the judgements that replace Aristotelian modes of inference will be presented in §22 (formulas 59, 62, 65).

Negation

§7. If a small vertical stroke is attached to the underside of the content stroke, this is to express the circumstance that *the content does not occur*. Thus, for example,

|—┬— *A*

means: "*A* does not occur." I call this small vertical stroke the *negation stroke*. The part of the horizontal stroke to the right of the negation stroke is the content stroke of *A*; while the part to the left of the negation stroke is the content stroke of the negation of *A*. Here, as in other places in the "conceptual notation", without the judgement stroke, no judgement is made.

calls upon us merely to form the idea that *A* does not occur, without expressing whether this idea is true.

We now consider some cases in which the symbols for conditionality and negation are combined.

---

* Editor's note: Frege here refers to Chapter II, which consists of several derivations within this system.

means: "The case in which *B* is to be affirmed and the negation of *A* is to be denied does not occur." In other words, "The possibility of affirming both *A* and *B* does not exist." Therefore, only the following three cases remain:

A is affirmed and *B* is denied
A is denied and *B* is affirmed
A is denied and *B* is denied.

In view of the preceding, it is easy to demonstrate the significance of each of the three parts of the horizontal stroke to the left of *A*.

means: "The case in which *A* is denied and the negation of *B* is affirmed does not obtain." or, "*A* and *B* cannot both be denied." Only the following possibilities remain:

A is affirmed and *B* is affirmed
A is affirmed and *B* is denied
A is denied and *B* is affirmed.

A and *B* together exhaust all the possibilities. Now the words "or" and "either—or" are used in two ways:

"*A* or *B*"

means, first, just the same as

$$\begin{array}{c} A \\ B; \end{array}$$

thus, that nothing is conceivable other than *A* and *B*. For example: If a quantity of gas is heated, its volume or its pressure increases. In its second use, the expression

"*A* or *B*"

unites the meanings of

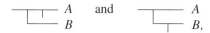

so that, first, there is no third possibility outside of *A* and *B*; and secondly, *A* and *B* are mutually exclusive. Of the [original] four possibilities, then, only the following two remain:

A is affirmed and *B* is denied

A is denied and *B* is affirmed.

Of the two uses for the expression "*A* or *B*", the first, in which the coexistence of *A* or *B* is not excluded, is the more important; and *we shall use the word "or" with this meaning.* Perhaps it is appropriate to make this distinction between "or" and "either-or" that only the latter shall have the secondary meaning of mutual exclusion. We can then translate

$$\begin{array}{l}\rule{1cm}{0.4pt}\!\!\top\!\!-A \\ \sqcup\!\!\top\!\!-B\end{array}$$

by "*A* or *B*". Similarly,

$$\begin{array}{l}\rule{1cm}{0.4pt}\!\!\top\!\!-A \\ \sqcup\!\!\top\!\!-B \\ \sqcup\!\!\top\!\!-\Gamma\end{array}$$

has the meaning of "*A* or *B* or Γ".

$$\vdash\!\!\top\!\!\top\!\!-A \\ \sqcup\!\!-B$$

means "$\rule{0.6cm}{0.4pt}\!\top\!\!\top\!\!-A \atop \sqcup\!\!-B$

is denied"; or, "the case in which *A* and *B* are both affirmed occurs". Conversely, the three possibilities left open by

$$\rule{1cm}{0.4pt}\!\!\top\!\!\top\!\!-A \\ \sqcup\!\!-B$$

are excluded. Accordingly, we can translate

$$\vdash\!\!\top\!\!\top\!\!-A \\ \sqcup\!\!-B$$

by: "Both *A* and *B* are facts." Also, we can easily see that

$$\top\!\!\top\!\!\top\!\!-A \\ \sqcup\!\!-B \\ \sqcup\!\!-\Gamma$$

can be rendered by: "*A* and *B* and Γ". If we wish to present "either *A* or *B* in symbols with the secondary meaning of mutual exclusion, we must express "$\rule{0.6cm}{0.4pt}\!\top\!\!\top\!\!-A \atop \sqcup\!\!-B$ and $\rule{0.6cm}{0.4pt}\!\top\!\!-A \atop \sqcup\!\!\top\!\!-B$." This gives

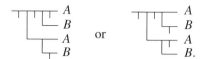

Instead of expressing "and" by means of the symbols for conditionality and negation, as is done here, we could, conversely, represent conditionality by means of a symbol for "and" and the symbol for negation. We could introduce, say,

$$\left\{\begin{array}{c} \varGamma \\ \varDelta \end{array}\right.$$

as the symbol for the combined content of $\varGamma$ and $\varDelta$, and then render

$$\begin{array}{c} \text{---}\!\top\!\!\begin{array}{c} A \\ B \end{array} \end{array}$$

by

$$\begin{array}{c} \top\!\left\{\top\!\!\begin{array}{c} A \\ B. \end{array}\right. \end{array}$$

I chose the other way because deduction seemed to me to be expressed more simply that way. The distinction between "and" and "but" is the kind that is not expressed in this "conceptual notation". The speaker uses "but" when he wishes to give a hint that what follows is different from what one might at first suppose.

$$\vdash\!\top\!\!\begin{array}{c} A \\ B \end{array}$$

means: "Of the four possibilities, the third, namely that A is denied and B affirmed, occurs." Thus, we can translate [this formula by]:

"B and (but) not A occurs."

We can translate the symbol combination

$$\vdash\!\top\!\!\begin{array}{c} B \\ A \end{array}$$

in the same way.

$$\vdash\!\top\!\!\begin{array}{c} B \\ A \end{array}$$

means: "The case in which A and B are both denied occurs." We can thus translate [this formula by]:

"Neither *A* nor *B* is a fact."

Obviously, the words "or", "and", "neither–nor" are here considered only in so far as they combine *assertible* contents.

Identity of content

§8. Identity of content differs from conditionality and negation by relating to names, not to contents. Although symbols are usually only representatives of their contents – so that each combination [of symbols usually] expresses only a relation between their contents – they at once appear *in propria persona* as soon as they are combined by the symbol for identity of content, for this signifies the circumstance that the two names have the same content. Thus, with the introduction of a symbol for identity of content, the bifurcation is necessarily introduced into the meaning of every symbol, the same symbols standing at times for their contents, at times for themselves. This gives the impression, at first, that what we are dealing with pertains merely to the *expression* and *not to the thought*, and that we have absolutely no need for different symbols of the same content, and thus no [need for a] symbol for identity of content either. To show the falsity of this appearance, I choose the following example from geometry: Let a straight line rotate about a fixed point *A* on the circumference of a circle. When the straight line forms a diameter [of the circle], let us call the end [of the diameter] opposite *A* the point *B* corresponding to this position [of the straight line]. Then let us go on to call the point of intersection of the two lines [that is, the circumference and the straight line] the point *B* corresponding to the position of the straight line at any given time. This [point *B*] is such that it follows the rule that continuous changes in the position of the straight line must always correspond to continuous changes in the position of *B*.

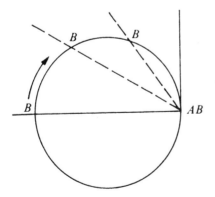

[As the line turns in the direction of the arrow, *B* moves towards *A*, till they coincide.]

Therefore, the name *B* denotes something undetermined as long as the corresponding position of the straight line is not yet specified. We can now ask: What point corresponds to the position of the straight line when it is perpendicular to the diameter? The answer will be: The point A. Thus, in this case, the name B has the same content as the name A; and yet we could not have used only one name from the beginning since the justification for doing so is first given by our answer. The same point is determined in two ways:

(1) Directly through perception.
(2) As the point B corresponding to the [rotating] straight line's being perpendicular to the diameter.

A separate name corresponds to each of these two modes of determination. Thus, the need of a symbol for identity of content rests upon the following fact: the same content can be fully determined in different ways; but, at the *same content*, in a particular case, is actually given by *two {different} modes of determination* is the content of a *judgement*. Before this [judgement] can be made, we must supply two different names, corresponding to the two [different] modes of determination, for the thing thus determined. But the judgement requires for its expression a symbol for identity of content to combine the two names. It follows from this that different names for the same content are not always merely an indifferent matter of form; but rather, if they are associated with different modes of determination, they concern the very heart of the matter. In this case, the judgement as to identify the content is, in Kant's sense synthetic. A more superficial reason for the introduction of a symbol for identity of content is that it is occasionally convenient to introduce an abbreviation for a lengthy expression. We must then express identity of content between the abbreviation and the original form.

Now, let $\qquad \vdash\!\!\!\!\!\!-\!\!-\!\!-\ (A\equiv B)$

mean: *the symbol A and the symbol B have the same conceptual content, so that we can always replace A by B and vice versa.*

The function

§9. Let us suppose that the circumstance that hydrogen is lighter than carbon dioxide is expressed in our formula language. Then, in place of the symbol for hydrogen, we can insert the symbol for oxygen or for nitrogen. By this means, the sense is altered in such a way that "oxygen" or "nitrogen" enters into the relations in which "hydrogen" stood before. If we think of an expression as variable in this way, it divides into [1] a constant component which represents the totality of the relations and [2] the symbol which is regarded as replaceable by others and which denotes the object which stands in these relations. I call the first component a function, the second its argument. This distinction has nothing to do with the conceptual content, but only with our way of viewing it. Although, in the mode of consideration just indicated, "hydrogen" was the argument and "being lighter than carbon dioxide" the function, we can also apprehend the same conceptual content in such a way that "carbon dioxide" becomes the argument and "being heavier than hydrogen" the function. In this case we need only think of "carbon dioxide" as replaceable by other ideas like "hydrogen chloride gas" or "ammonia".

"The circumstance that carbon dioxide is heavier than hydrogen"
and
"the circumstance that carbon dioxide is heavier than oxygen"

are the same function with different arguments if we regard "hydrogen" and "oxygen" as arguments. On the other hand, they are different functions of the same argument if we consider "carbon dioxide" as the argument.

Consider now as our example: "the circumstance that the centre of mass of the solar system has no acceleration, if only internal forces act on the solar system". Here "solar system" occurs in two places. We can thus consider this as a function of the argument "solar system" in various ways, according as we think of "solar system" as replaceable by other arguments at the first or the second or at both places – but in the last case, replaceable by the same thing both times. These three functions are all different. The proposition that Cato killed Cato shows the same thing. Here, if we think of "Cato" as replaceable at the first occurrence, then "killing Cato" is the function. If we think of "Cato" as replaceable at the second occurrence, then "being killed by Cato" is the function. Finally, if we think of "Cato" as replaceable at both occurrences, then "killing oneself" is the function.

We now express the matter generally:

If, in an expression (whose content need not be assertible), a simple or complex symbol occurs in one or more places and we imagine it as replaceable by another [symbol] (but the same one each time) at all or some of these places, then we call the part of the expression that shows itself invariant [under such replacement] a function and the replaceable part its argument.

Since, according to this account, something can occur as an argument and also at places in the function where it is not regarded as replaceable, we distinguish between argument-places and other places in the function.

Let me warn here against an illusion to which the use of [ordinary] language easily gives rise. If we compare the two propositions:

"The number 20 can be represented as the sum of four squares."
and
"Every positive integer can be represented as the sum of four squares."

it appears possible to consider "being representable as the sum of four squares" as a function whose argument is "the number 20" one time, and "every positive integer" the other time. We can discern the error of this view from the observation that "the number 20" and "every positive integer" are not concepts of the same rank. What is asserted of the number 20 cannot be asserted in the same sense of [the concept] "every positive integer"; though, of course, in some circumstances it may be asserted of every positive integer. The expression "every positive integer" by itself, unlike [the expression] "the number 20", yields no independent idea; it acquires a sense only in the context of a sentence.

For us, the different ways in which the same conceptual content can be considered as a function of this or that argument have no importance so long as function and argument are completely determinate. But if the argument becomes *indeterminate*, as in the judgement: "Whatever arbitrary positive integer we take as argument for 'being representable as the sum of four squares', the [resulting] proposition is always true.", then the distinction between function and argument acquires a *substantive* significance. It can also happen that, conversely, the argument is determinate, but the function is indeterminate. In both cases, through the opposition of the *determinate* and the *indeterminate* or the *more* and the *less determinate*, the whole splits up into *function* and *argument* according to its own content, and not just according to our way of looking at it.

If we imagine that in a function a symbol, which has so far been regarded as not replaceable,* is now replaceable at some or all of the places where it occurs, we then obtain, by considering it in this way, a function with another argument besides the one it had before. In this way, *functions of two or more arguments* arise. Thus, for example, "the circumstance that hydrogen is lighter than carbon dioxide" can be considered a function of the two arguments "hydrogen" and "carbon dioxide".

In the mind of the speaker, the subject is usually the principal argument; the next in importance often appears as the object. [Ordinary] language, through the choice of [grammatical] forms or of words, for example,

active–passive   heavier–lighter   give–receive

is free to permit, at will, this or that component of the sentence to appear as the principal argument; however, this freedom is limited by the scarcity of words.

§10. *In order to express an indeterminate function of the argument A, we put A in parentheses following a letter*, for example:

$$\Phi(A).$$

Similarly $\Psi(A,B)$

---

* We can also consider a symbol already previously regarded as replaceable [at some places] as now further replaceable at those positions where it was previously considered constant.

represents a function (not more explicitly determined) of the two arguments A and B. Here, the places of A and B in the parentheses represent the positions that A and B occupy in the function, regardless of whether *A* or *B* each occupies one place or more [in that function]. *Thus, in general,*

$$\Psi(A,B) \text{ and } \Psi(B,A)$$

are different.

Indeterminate functions of more [than two] arguments are expressed in a corresponding way.
We can read $\qquad \vdash\!\!\!\!\!-\!\!\!- \Phi(A)$

as: *A* has the property Φ."

$$\vdash\!\!\!\!\!-\!\!\!- \Psi(A,B)$$

can be translated by "*B* stands in the Ψ-relation to *A*." or "*B* is a result of an application of the procedure Ψ to the object *A*."
Since the symbol Φ occurs at a place in the expression

$$\Phi(A)$$

and since we can think of it as replaced by other symbols [such as] Ψ, X—which then express other functions of the argument *A*—we can consider $\Phi(A)$ as a function of the argument Φ. This shows quite clearly that the concept of function in analysis, which I have in general followed, is far more restricted than the one developed here.

Generality

§11. In the expression of a judgement we can always regard the combination of symbols to the right of $\vdash\!\!\!\!\!-\!\!\!-$ as a function of one of the symbols occurring in it. *If we replace this argument by a German letter and introduce in the content stroke a concavity containing the same German letter, as in*

$$\vdash\!\!\!-\!\!\stackrel{\frown{\mathfrak{a}}}{}\!\!-\!\!- \Phi(\mathfrak{a})$$

then this stands for the judgement that the function is a fact whatever we may take as its argument. Since a letter which is used as a function symbol, like Φ in $\Phi(A)$, can itself be considered as the argument of a function, it can be replaced by a German letter in the manner just specified. The meaning of a German letter is subject only to the obvious restrictions that [1] the assertibility (§2) of a combination of symbols following the content stroke must remain intact, and [2] if the German letter appears as a function symbol, this circumstance must be taken into account. *All other conditions that must be imposed upon what may be put in for a German letter are to be included in the judgement.* Thus, from such a judgement, we can always derive an arbitrary number of *judgements with less general content* by putting something different each time in place of the German letter; when we do this, the concavity in the content stroke disappears again.

The horizontal stroke situated left of the concavity in

$$\vdash\!\!-\!\!\smile^{\mathfrak{a}}\!\!-\!\!-\ \varPhi(\mathfrak{a})$$

is the content stroke of [the assertible content] that $\varPhi(\mathfrak{a})$ holds, whatever we may put in the place of $\mathfrak{a}$. The horizontal stroke to the right of the concavity is the content stroke of $\varPhi(\mathfrak{a})$, and here we must think of $\mathfrak{a}$ as replaced by something definite.

After what has been said above about the significance of the judgement stroke, it is easy to see what an expression like

$$-\!\!-\!\!\smile^{\mathfrak{a}}\!\!-\!\!-\ X(\mathfrak{a})$$

means. This can occur as part of a judgement like

$$\vdash\!\!-\!\!\smile^{\mathfrak{a}}\!\!-\!\!-\ X(\mathfrak{a}), \qquad \vdash\!\!-\!\!\begin{array}{c}\!\!-\!\!-\ A\\ \!\!\smile^{\mathfrak{a}}\!\!-\!\!-\ X(\mathfrak{a}).\end{array}$$

It is obvious that, unlike

$$\vdash\!\!-\!\!\smile^{\mathfrak{a}}\!\!-\!\!-\ \varPhi(\mathfrak{a})$$

these judgements cannot be used to derive less general judgements by replacing $\mathfrak{a}$ by something definite. By means of $\vdash\!\!-\!\!\smile^{\mathfrak{a}}\!\!-\!\!- X(\mathfrak{a})$, it is denied that $X(\mathfrak{a})$ is always a fact whatever we may put in place of $\mathfrak{a}$. This in no way denies that we can specify a meaning $\varDelta$ such that $X(\varDelta)$ would be a fact.

$$\vdash\!\!-\!\!\begin{array}{c}\!\!-\!\!-\ A\\ \!\!\smile^{\mathfrak{a}}\!\!-\!\!-\ X(\mathfrak{a})\end{array}$$

means that the case in which $-\!\!\smile^{\mathfrak{a}}\!\!- X(\mathfrak{a})$ is affirmed and A is denied does not occur. But this in no way denies the occurrence of the case in which $X(\varDelta)$ is affirmed and A denied; for as we have just seen, $X(\varDelta)$ can be affirmed and yet $-\!\!\smile^{\mathfrak{a}}\!\!-\ X(\mathfrak{a})$ denied. Thus, here also we cannot arbitrarily substitute for $\mathfrak{a}$ without jeopardizing the truth of the judgement. This explains why the concavity with the German letter written in it is necessary: *it delimits the scope of the generality signified by the letter. The German letter retains its significance only within its scope.* The same German letter can occur in various scopes in one judgement without the meaning that we may ascribe to it in one scope extending to [any of] the other scopes. The scope of one German letter can include that of another, as is shown by the example

$$\vdash\!\!\overset{\frown}{\underset{\underset{e}{\frown}}{\alpha}}\!\!\begin{array}{l}A(\mathfrak{a})\\ B(\mathfrak{a},\mathfrak{e}).\end{array}$$

In this case, *different* letters must be chosen; we may not replace 𝔢 by 𝔞. Naturally, it is permitted to replace one German letter throughout its scope by another particular one provided that there are still different letters standing where different letters stood before. This has no effect on the content. *Other substitutions are permitted only if the concavity follows immediately after the judgement stroke* so that the content of the whole judgement constitutes the scope of the German letter. Since, accordingly, this is a specially important case, I shall introduce the following abbreviation for it: *An italic letter is always to have as its scope the content of the whole judgement*, and this need not be signified by a concavity in the content stroke. If an italic letter occurs in an expression which is not preceded by a judgement stroke then this expression has no sense. *An italic letter may always be replaced by a German letter which does not yet occur in the judgement*; when this is done, the concavity must be placed immediately after the judgement stroke. For example, instead of

$$\vdash\!\!\!\!\!-\!\!\!\!- X(a)$$

we may put $\quad\vdash\!\!\overset{\frown}{\underset{}{\alpha}}\!\!- X(\mathfrak{a})$

if *a* only occurs in the argument places of *X(a)*.

It is also obvious that from

$$\vdash\!\!\overline{}\begin{array}{l}\Phi(a)\\ A\end{array}$$

we can derive $\quad\vdash\!\!\overset{\frown}{\underset{}{\alpha}}\!\!\overline{}\begin{array}{l}\Phi(\mathfrak{a})\\ A\end{array}$

if A is an expression in which a does not occur and a stands only in argument places of Φ(a). If $\overset{\frown}{\underset{}{\alpha}}\!\!-\Phi(\mathfrak{a})$ is denied, then we must be able to specify a meaning for *a* such that *Φ(a)* is denied. Thus, if $\overset{\frown}{\underset{}{\alpha}}\!\!-\Phi(\mathfrak{a})$ were denied and *A* affirmed, then we should have to be able to specify a meaning for *a* such that *A* would be affirmed and *Φ(a)* denied. But because of

$$\vdash \begin{array}{l} \Phi(a) \\ A \end{array}$$

we cannot do this; for this [formula] means that, whatever a may be, the case in which $\Phi(a)$ would be denied and A affirmed is excluded. Thus, we cannot deny 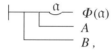 $\Phi(\mathfrak{a})$ and affirm A; that is:

$$\vdash \begin{array}{l} \mathfrak{a} \\ \Phi(\mathfrak{a}) \\ A \,. \end{array}$$

Similarly, from

$$\vdash \begin{array}{l} \Phi(a) \\ A \\ B \,, \end{array}$$

we can deduce

$$\vdash \begin{array}{l} \mathfrak{a} \\ \Phi(\mathfrak{a}) \\ A \\ B \,, \end{array}$$

if a does not occur in A or B and $\Phi(a)$ contains a only in argument places. This case can be reduced to the preceding one, since instead of

$$\vdash \begin{array}{l} \Phi(a) \\ A \\ B \end{array}$$

we can put

$$\vdash \begin{array}{l} \Phi(a) \\ A \\ B \,, \end{array}$$

and

$$\vdash \begin{array}{l} \mathfrak{a} \\ \Phi(\mathfrak{a}) \\ A \\ B \end{array}$$

can be converted again into

$$\vdash \begin{array}{l} \mathfrak{a} \\ \Phi(\mathfrak{a}) \\ A \\ B. \end{array}$$

A similar treatment holds when there are still more conditional strokes.

§12. We now consider some combinations of symbols.

$$\vdash\!\!\!\top\!\!-\!\!\stackrel{\alpha}{\smile}\!\!-X(\alpha)$$

means that we can find something, say Δ, such that $X(\Delta)$ would be denied. We can thus translate it: "There are some things which do not have the property X."

The sense of

$$\vdash\!\!-\!\stackrel{\alpha}{\smile}\!\!\top\!\!-X(\alpha)$$

is different. This means: "Whatever α may be, $X(\alpha)$ must always be denied."; or: "There does not exist something having the property X."; or, if we call something that has the property X a X: "There is no X."

$-\!\stackrel{\alpha}{\smile}\!\!\top\!\!- \Lambda(\alpha)$ is denied by

$$\vdash\!\!\!\top\!\!-\!\!\stackrel{\alpha}{\smile}\!\!\top\!\!-\Lambda(\alpha).$$

Thus, we can translate it: "There are Λ's."*

$$\vdash\!\!-\!\stackrel{\alpha}{\smile}\!\!\top\!\!\begin{array}{c}- P(\alpha) \\ \hphantom{-} \\ \hline - X(\alpha) \end{array}$$

means: "Whatever may be put in place of α, the case in which $P(\alpha)$ would have to be denied and $X(\alpha)$ affirmed does not occur." It is thus possible here that for some meanings which we could give to α,

$P(\alpha)$ would have to be affirmed and $X(\alpha)$ affirmed,
for others
$P(\alpha)$ would have to be affirmed and $X(\alpha)$ denied,
for still others
$P(\alpha)$ would have to be denied and $X(\alpha)$ denied.

We can thus translate: "If something has the property X, then it also has the property P.", or "Every X is a P.", or "All Xs are Ps." *This is how causal connections are expressed.*

$$\vdash\!\!-\!\stackrel{\alpha}{\smile}\!\!\top\!\!\begin{array}{c}- P(\alpha) \\ \hline - \Psi(\alpha) \end{array}$$

---

* This must be understood to include the case "There exists one Λ." For example, if $\Lambda(x)$ signifies the circumstance that x is a house, then

$$\vdash\!\!\!\top\!\!-\!\!\stackrel{\alpha}{\smile}\!\!\top\!\!- \Lambda(\alpha)$$

means: "There are houses or is at least one house." See §2, second footnote.

means: "A meaning cannot be given to α such that $P(α)$ and $\Psi(α)$ could both be affirmed." Thus, we could translate: "What has the property Ψ does not have the property P", or "No Ψ is a P."

denies
$$\vdash\!\!\!\!\overset{α}{\frown}\!\!\!\!\begin{array}{l} P(α) \\ \Lambda(α) \end{array}$$

denies $\overset{α}{\frown}\!\!\!\!\begin{array}{l} P(α) \\ \Lambda(α) \end{array}$ and can thus be rendered as "Some Λ's are not P's."

$$\vdash\!\!\!\!\overset{α}{\frown}\!\!\!\!\top\!\!\!\!\begin{array}{l} P(α) \\ M(α) \end{array}$$

denies that no M is a P and thus means: "Some* M's are P's.", or "It is possible for a M to be a P."

Thus we obtain the square of logical opposition:

$\overset{α}{\frown}\!\!\!\!\begin{array}{l} P(α) \\ X(α) \end{array}$ CONTRARY $\overset{α}{\frown}\!\!\!\!\top\!\!\!\!\begin{array}{l} P(α) \\ X(α) \end{array}$

```
S                 C                               Y          S
U                   O                           R            U
B                     N                       O              B
A                       T                   T                A
L                         R               C                  L
T                           A           I                    T
E                             D                              E
R                           A           I                    R
N                         R               C                  N
A                       T                   T                A
T                     N                       O              T
E                   O                           R            E
                  C                               Y
```

$\top\!\!\!\!\overset{α}{\frown}\!\!\!\!\top\!\!\!\!\begin{array}{l} P(α) \\ X(α) \end{array}$ [SUB] CONTRARY $\top\!\!\!\!\overset{α}{\frown}\!\!\!\!\top\!\!\!\!\begin{array}{l} P(α) \\ X(α) \end{array}$

---

* The word "some" must always be understood here to include the case "one". In a more lengthy manner, we could say: "some, or at least one."

2
ON THE SCIENTIFIC JUSTIFICATION OF A CONCEPTUAL NOTATION

Time and again, in the more abstract regions of science, the lack of a means of avoiding misunderstandings on the part of others, and also errors in one's own thought, makes itself felt. Both [shortcomings] have their origin in the imperfection of language, for we do have to use sensible symbols to think.

Our attention is directed by nature to the outside. The vivacity of sense-impressions surpasses that of memory-images to such an extent that, at first, sense-impressions determine almost by themselves the course of our ideas, as is the case in animals. And we would scarcely ever be able to escape this dependency if the outer world were not to some extent dependent upon us.

Even most animals, through their ability to move about, have an influence on their sense-impressions: they can flee some, seek others. And they can even effect changes in things. Now man has this ability to a much greater degree; but nevertheless, the course of our ideas would still not gain its full freedom from this [ability alone]: it would still be limited to that which our hand can fashion, our voice intone, without the great invention of symbols which call to mind that which is absent, invisible, perhaps even beyond the senses.

I do not deny that even without symbols the perception of a thing can gather about itself a group of memory-images; but we could not pursue these further: a new perception would let these images sink into darkness and allow others to emerge. But if we produce the symbol of an idea which a perception has called to mind, we create in this way a firm, new focus about which ideas gather. We then select another [idea] from these in order to elicit *its* symbol. Thus we penetrate step by step into the inner world of our ideas and move about there at will, using the realm of sensibles itself to free ourselves from its constraint. Symbols have the same importance for thought that discovering how to use the wind to sail against the wind had for navigation. Thus, let no one despise symbols! A great deal depends upon choosing them properly. And their value is not diminished by the fact that, after long practice, we need no longer produce [external] symbols, we need no longer speak out loud in order to think; for we think in words nevertheless, and if not in words, then in mathematical or other symbols.

Also, without symbols we would scarcely lift ourselves to conceptual thinking. Thus, in applying the same symbol to different but similar things, we actually no longer symbolize the individual thing, but rather what [the similars] have in common: the concept. This concept is first gained by symbolizing it; for since it is, in itself, imperceptible, it requires a perceptible representative in order to appear to us.

This does not exhaust the merits of symbols; but it may suffice to demonstrate their indispensability. Language proves to be deficient, however, when it comes to protecting thought from error. It does not even meet the first requirement which we must place upon it in this respect; namely, being unambiguous. The most dangerous cases [of ambiguity] are those in which the meanings of a word are only slightly different, the subtle and yet not unimportant variations. Of the many examples [of this kind of ambiguity] only one frequently recurring phenomenon may be mentioned here: the same word may serve to designate a concept and a single object which falls under that concept. Generally, no strong distinction is made between concept and individual. "The horse" can denote a single creature; it can also denote the species, as in the sentence: "The horse is an herbivorous animal." Finally, horse can denote a concept, as in the sentence: "This is a horse."

Language is not governed by logical laws in such a way that mere adherence to grammar would guarantee the formal correctness of thought processes. The forms in which inference is expressed are so varied, so loose and vague, that presuppositions can easily slip in unnoticed and then be overlooked when the necessary conditions for the conclusion are enumerated. In this way, the conclusion obtains a greater generality than it justifiably deserves.

Even such a conscientious and rigorous writer as Euclid often makes tacit use of presuppositions which he specifies neither in his axioms and postulates nor in the premisses of the particular theorem [being proved]. Thus, in the proof of the nineteenth theorem of the first book of *The Elements* (in every triangle, the largest angle lies opposite the largest side), he tacitly uses the statements:

(1) If a line segment is not larger than a second one, the former is equal to or smaller than the latter.
(2) If an angle is the same size as a second one, the former is not larger than the latter.
(3) If an angle is smaller than a second one, the former is not larger than the latter.

Only by paying particular attention, however, can the reader become aware of the omission of these sentences, especially since they seem so close to being as fundamental as the laws of thought that they are used just like those laws themselves.

A strictly defined group of modes of inference is simply not present in [ordinary] language, so that on the basis of linguistic form we cannot distinguish between a "gapless" advance [in the argument] and an omission of connecting links. We can even say that the former almost never occurs in [ordinary] language, that it runs against the feel of language because it would involve an insufferable prolixity. In [ordinary] language, logical relations are almost always only hinted at — left to guessing, not actually expressed.

The only advantage that written word has over the spoken word is permanence: [with the written word], we can review a train of thought many times without fear that it will change; and thus we can test its validity more thoroughly. In this [process of testing], since insufficient security lies in the nature of the word-language itself, the laws of logic are applied externally like a plumb-line. But even so, mistakes easily escape the eye of the examiner, especially those which arise from subtle differences in the meanings of a word. That we nevertheless find our way about reasonably well in life as well as in science we owe to the manifold ways of checking that we have at our disposal. Experience and space perception protect us from many errors. Logical rules [, externally applied,] furnish little protection, as is shown by examples from disciplines in which the means of checking begin to fail. These rules have failed to defend even great philosophers from mistakes, and have helped just a little in keeping higher mathematics free from error, because they have always remained external to content.

The shortcomings stressed [here] are rooted in a certain softness and instability of [ordinary] language, which nevertheless is necessary for its versatility and potential for development. In this respect [ordinary] language can be compared to the hand, which despite its adaptability to the most diverse tasks is still inadequate. We build for ourselves artificial hands, tools for particular purposes, which work with more accuracy than the hand can provide. And how is this accuracy possible? Through the very stiffness and inflexibility of parts the lack of which makes the hand so dextrous. Word-language is inadequate in a similar way. We need a system of symbols from which every ambiguity is banned, which has a strict logical form from which the content cannot escape.

We may now ask which is preferable, audible symbols or visible ones. The former have, first of all, the advantage that their production is more independent of external circumstances. Furthermore, much can be made in particular of the close kinship of sounds to inner processes. Even their form of appearance, the temporal sequence, is the same; both are equally fleeting. In particular, sounds have a more intimate relation to the emotions than shapes and colours do; and the human voice with its boundless flexibility is able to do justice to even the most delicate combinations and variations of feelings. But no matter how valuable these advantages may be for other purposes, they have no importance for the rigour of logical deductions. Perhaps this intimate adaptability of audible symbols to the physical and mental conditions of reason has just the disadvantage of keeping reason more dependent upon these.

It is completely different with visible things, especially shapes. They are generally sharply defined and clearly distinguished. This definiteness of written symbols will tend to make what is signified also more sharply defined; and just such an effect upon ideas must be asked for the rigour of deduction. This can be achieved, however, only if the symbol directly denotes the thing [symbolized].

A further advantage of the written symbol is greater permanence and immutability. In this way, it is also similar to the concept—as it should be—and thus, of course, the more dissimilar to the restless flow of our actual thought processes. Written symbols offer the possibility of keeping many things in mind at the same time; and even if, at each moment, we can only concentrate upon a small part of these, we still retain a general impression of what remains, and this is immediately at our disposal whenever we need it.

The spatial relations of written symbols on a two-dimensional writing surface can be employed in far more diverse ways to express inner relations than the mere following and preceding in one-dimensional time, and this facilitates the apprehension of that to which we wish to direct our attention. In fact, simple sequential ordering in no way corresponds to the diversity of logical relations through which thoughts are interconnected.

Thus, the very properties which set the written symbol further apart [than the spoken word] from the course of our ideas are most suited to remedy certain shortcomings of our make-up. Therefore, when it is not a question of representing natural thought as it actually took shape in reciprocal action with the word-language, but concerns instead the supplementation of the onesidedness of thinking which results from a close connection with the sense of hearing, then the written symbol will be preferable. Such a notation must be completely different from all word-languages in order to exploit the peculiar advantages of written symbols. It need hardly be mentioned that these advantages scarcely come into play at all in the written word. The relative position of the words with respect to each other on the writing surface depends to a large extent upon the length of the lines [of print] and is, thus, without importance. There are, however, completely different kinds of notation which better exploit these [mentioned] advantages. The arithmetic language of formulas is a conceptual notation since it directly expresses the facts without the intervention of speech. As such, it attains a brevity which allows it to accommodate the content of a simple judgement in one line. Such contents — here equations or inequalities — as they follow from one another are written under one another. If a third follows from two others, we separate the third from the first two with a horizontal stroke, which can be read "therefore". In this way, the two-dimensionality of the writing surface is utilized for the sake of perspicuity. Here the deduction is stereotyped being almost always based upon identical transformations of identical numbers yielding identical results. Of course, this is by no means the only method of inference in arithmetic; but where the logical progression is different, it is generally necessary to express it in words. Thus, the arithmetic language of formulas lacks expressions for logical connections; and, therefore, it does not merit the name of conceptual notation in the full sense.

Exactly the opposite holds for the symbolism for logical relations originating with Leibniz and revived in modern times by Boole, R. Grassmann, S. Jevons, E. Schröder, and others. Here we do have the logical forms, though not entirely complete; but content is lacking. In these cases, any attempt to replace the single letters with expressions of contents, such as analytic equations, would demonstrate with the resulting imperspicuity, clumsiness — even ambiguity — of the formulas how little suited this kind of symbolism is for the construction of a true conceptual notation.

I would demand the following from a true conceptual notation: It must have simple modes of expression for the logical relations which, limited to the necessary, can be easily and surely mastered. These forms must be suitable for combining most intimately with a content. Also, such brevity must be sought that the two-dimensionality of the writing surface can be exploited for the sake of perspicuity. The symbols for denoting content are less essential. They can be easily created as required, once the general [logical] forms are available. If the analysis of a concept into its ultimate components does not succeed or appears unnecessary, we can be content with temporary symbols.

It would be easy to worry unnecessarily about the feasibility of the matter. It is impossible, someone might say, to advance science with a conceptual notation, for the invention of the latter already presupposes the completion of the former. Exactly the same apparent difficulty arises for [ordinary] language. This is supposed to have made reason possible, but how could man have invented language without reason? Research into the laws of nature employs physical instruments; but these can be produced only by means of an advanced technology, which again is based upon knowledge of the laws of nature. The [apparently vicious] circle is resolved in each case in the same way: an advance in physics results in an advance in technology, and this makes possible the construction of new instruments by means of which physics is again advanced. The application [of this example] to our case is obvious.

Now I have attempted* to supplement the formula language of arithmetic with symbols for the logical relations in order to produce — at first just for arithmetic — a conceptual notation of the kind I have presented as desirable. This does not rule out the application of my symbols to other fields. The logical relations occur everywhere, and the symbols for particular contents can be so chosen that they fit the framework of the conceptual notation. Be that as it may, a perspicuous representation of the forms of thought has, in any case, significance extending beyond mathematics. May philosophers, then, give some attention to the matter!

---

* *Conceptual Notation, a Formula Language of Pure Thought modelled upon the Formula Language of Arithmetic*, Halle a. S., 1879.

3
ON THE AIM OF THE "CONCEPTUAL NOTATION"

I had the honour once before to give a paper here about my "conceptual notation". What induces me to return to it again is the observation that its aim has frequently been misunderstood. This I gather from several reviews of my book which have appeared since [I gave my last paper here]. Distorted judgements must have resulted from these reviews.

I have been rebuked for failing, among other things, to consider the achievements of Boole. Among those who make this reproach is E. Schröder in the review in the *Zeitschrift für Mathematik und Physik*, XXV. In comparing my "conceptual notation" with the Boolean formula language, he reaches the conclusion that the latter is preferable in every respect. Although this judgement can hardly please me, I am still thankful to him for the detailed review and the technical reasons for his objections, since they give me the opportunity to refute the objections and set the matter in a brighter light.

With respect to the reproach just mentioned, I wish to say first of all that The Boolean formula language, in the more than twenty years which have passed since its invention, has in no way had such a smashing success that leaving out the foundation it established would, as a matter of course, necessarily appear foolish, and that only a further development could be considered. Besides, the problems which Boole treats appear to be, for the most part, first fabricated in order to be solved with his formulas.

This reproach, however, essentially overlooks the fact that my aim was different from Boole's. I did not wish to present an abstract logic in formulas, but to express a content through written symbols in a more precise and perspicuous way than is possible with words. In fact, I wished to produce, not a mere *calculus ratiocinator*, but a *lingua characteristica* in the Leibnizian sense. In doing so, however, I recognize that deductive calculus is a necessary part of a conceptual notation. If this was misunderstood, perhaps it is because I let the abstract logical aspect stand too much in the foreground.

Now, in order to demonstrate in detail the differences in Boole's formula language and mine, I shall give first a short presentation of the former. We cannot consider all the variants which are found in Boole's predecessors and successors, since, compared to the considerable difference between my "conceptual notation" [and Boole's calculus], these others are not important.

Boole distinguishes *primary propositions* from *secondary propositions*. The former compare the extensions of concepts, the latter express relations among assertible contents. This division is insufficient since existential judgements fail to find a place.

Let us consider first *primary propositions*. Here the letters denote the extensions of concepts. Particular things as such are not signified; and this is an important deficiency in the Boolean formula language, for even if a concept covers only a single thing, a great difference still remains between it and this thing. Now the letters are combined with each other by logical multiplication and addition. If A means the extension of the concept "triangle" and B means the extension of the concept "equilateral", then the logical product

$$A.B$$

signifies the extension of the concept

$$A+B$$

is to be understood as the extension of the concept "triangle or equilateral".* The [use of the] expressions "product" and "sum" is justified by the existence of the following [correct] equations:

$$\begin{aligned} A.B &= B.A \\ A(B.C) &= (A.B).C \\ A+B &= B+A \\ A+(B+C) &= (A+B)+C \\ A(B+C) &= AB+AC \end{aligned}$$

But these points of agreement with algebraic multiplication and addition stand opposed to large discrepancies. The following are logically sound:

$$\begin{aligned} A &= A.A = A.A.A \\ A &= A+A = A+A+A \end{aligned}$$

but they do not generally hold in algebra. The differences in logical and mathematical calculation have so many important consequences that solving logical equations, with which Boole principally occupies himself, has scarcely anything in common with solving algebraic equations.

---

* Boole presupposes by this that the concepts A and B exclude each other, which, among others, Schröder does not.

Now the subordination of one concept to another can be expressed [in Boole's notation] this way:

$$A = A \cdot B.$$

For example, if A signifies the extension of the concept "mammal" and B signifies the extension of the concept "air-breathing", then the equation says: the extensions of the concepts "mammal" and "air-breathing mammal" are the same; that is, all mammals are air-breathing. The falling of an individual under a concept, which is totally different from the subordination of one concept to another, has no particular expression in Boole['s notation]; strictly speaking, none at all.

Everything thus far is already found with only superficial divergences in Leibniz, whose work in this area I dare say were unknown to Boole.

0 signifies for Boole the extension of a concept under which nothing falls; 1 means the extension of a concept under which everything that is being considered (*universe of discourse*) falls. We can see that here also the meaning of these symbols, especially of 1, deviates from arithmetic. Instead of these, Leibniz has "*non ens*" and "*ens*".

$$A \cdot B = 0$$

says that the two concepts exclude each other; like, for example, "the square-root of 2" and "whole number". The equation can hold without either

$$A + 0 \quad \text{or} \quad B = 0.$$

We still need a symbol for negation other than zero in order to convert, for example, the concept "man" into the concept "not man". Writers differ from each other here. For this purpose, Schröder attaches to the letters the index 1. Others have as well a symbol for the negation of identity. I do not consider this multiplicity of negation symbols an advantage of Boolean logic.

Boole reduces *secondary propositions* – for example, hypothetical and disjunctive judgements – to *primary propositions* in a very artificial way. He interprets the judgement "if $x = 2$, then $x^2 = 4$" this way: the class of moments of time in which $x = 2$ is subordinate to the class of moments of time in which $x^2 = 4$. Thus, here again the matter amounts to the comparison of the extensions of concepts; only here these concepts are fixed more precisely as classes of moments of time in which a sentence is true. This conception has the disadvantage that time becomes involved where it should remain completely out of the matter. MacColl explains the expressions for *secondary propositions* independently of the *primary* ones. In this way, the intermingling of time is certainly avoided; but as a result, every interconnection is severed between the two parts which, according to Boole, compose logic. We proceed, then, either in *primary propositions* and use the formulas in the sense stipulated by Boole; or else, we proceed in *secondary propositions* and use the interpretations of MacColl. Any [logical] transition from one kind of judgement to the other – which to be sure, often occurs in actual thinking – is blocked; for we may not use the same symbols with a double meaning in the same context.

When we view the Boolean formula language as a whole, we discover that it is a clothing of abstract logic in the dress of algebraic symbols. It is not suited for the rendering of a content, and that is also not its purpose. But this is exactly my intention. I wish to blend together the few symbols which I introduce and the symbols already available in mathematics to form a single formula language. In it, the existing symbols [of mathematics] correspond to the word-stems of [ordinary] language; while the symbols I add to them are comparable to the suffixes and [deductive] formwords that logically interrelate the contents embedded in the stems.

For this purpose, I could not use the Boolean symbolism; for it is not feasible to have, for example, the + sign occurring in the same formula part of the time in the logical sense and part of the time in the arithmetical sense. The analogy between the logical and arithmetical methods of calculation, which is of value to Boole, can only bring about confusion if both are combined together. Boole's symbolic language is conceivable only in complete separation from arithmetic. Therefore, I must invent other symbols for the logical relations.

Schröder says that my "conceptual notation" has almost nothing in common with Boole's calculus of concepts, but it does with his calculus of judgements. In fact, it is one of the most important differences between my mode of interpretation and the Boolean mode – and indeed I can add the Aristotelian mode – that I do not proceed from concepts, but from judgements. But this is certainly not to say that I would not be able to express the relation of subordination between concepts.

In front of an expression for an assertible content, such as $2 + 3 = 5$, I put a horizontal stroke, the content stroke, distinguishable from the minus sign by its greater length:

$$\vdash 2 + 3 = 5$$

I take this stroke to mean that the content which follows it is unified, so that other symbols can be related to it [as a whole].

In $\qquad \vdash 2 + 3 = 5$

absolutely no judgement is made. Thus, we can also write:

$$\vdash 4 + 2 = 7$$

without being guilty of writing a falsehood.

If I wish to assert a content as correct, I put the judgement stroke on the left end of the content stroke:

$$\vdash \quad 2 + 3 = 5$$

How thoroughly one is misunderstood sometimes! Through this mode of notation I meant to have a very clear distinction between the act of judging and the formation of a mere assertible content; and Rabus* accuses me of mixing up the two!

In order to express the negation of a content, I attach the negation stroke to the content stroke; for example

$$─\!\!\top\!\!─ \quad 4 + 2 = 7$$

The falsehood of this equation, however, is not asserted in this way. A new assertible content has been formed which becomes the judgement "4 + 2 does not equal 7" only by adding the judgement stroke:

$$\vdash\!\top\!─ \quad 4 + 2 = 7$$

If we wish to relate two assertible contents, A and B, to each other, we must consider the following cases:

(1) A and B

(2) A and not B

(3) not A and B

(4) not A and not B.

Now I understand by $\quad \top\!\!\!\begin{array}{l} A \\ B \end{array}$

the negation of the third case. This stipulation may appear very artificial at first. It is not clear at first why I single out this third case in particular and express its negation by a special symbol. The reason, however, will be immediately evident from an example:

$$\vdash\!\top\!\!\!\begin{array}{l} x^2 = 4 \\ x + 2 = 4 \end{array}$$

---

* *Die neuesten Bestrebungen auf dem Gebiete der Logik bei den Deutschen und die logische Frage,* Erlangen, 1880.

130   Logicism and the Philosophy of Language

denies the case that x^2 is not equal to 4 while nevertheless $x + 2$ is equal to 4. We can translate it: if $x + 2 = 4$, then $x^2 = 4$. This translation reveals the importance of the relation embedded in our symbol. Indeed, the hypothetical judgement is the form of all laws of nature and of all causal connections in general. To be sure, a rendering by means of "if" is not appropriate in all cases of linguistic usage, but only if an indeterminate constituent – like x here – confers generality on the whole. Were we to replace x by 2, then one would not appropriately translate

$$\vdash \begin{array}{l} 2^2 = 4 \\ 2 + 2 = 4 \end{array}$$

by "If $2 + 2 = 4$, then $2^2 = 4$".

Now consider the combinations of conditional stroke and negation stroke in the following table:

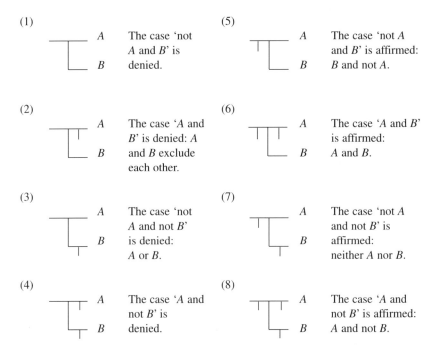

If we attach the negation stroke to the content strokes of the expressions on the left, we obtain the expressions next to them on the right. The denied case on the left is always affirmed on the right. The second expression arises from the first one by replacing A by the denial of A. Then in the verbal expression, the two denials of A cancel. The third expression arises from the first and the fourth from the second by the conversion of B into the denial of B. The "or" in the third case is the non-exclusive one. The exclusive "or" can be expressed this way:

I pause here to answer some objections of Schröder. He compares my representation of the exclusive "A or B" with his mode of writing

$$ab_1 + a_1b = 1$$

and finds here, as elsewhere in my "conceptual notation", a monstrous waste of space. In fact, I cannot deny that my expression takes up more room that Schröder's, which for its part is again more spreadout than Boole's original

$$a + b = 1.$$

But this criticism is based upon the view that my "conceptual notation" is supposed to be a presentation of abstract logic. These formulas [of my "conceptual notation"] are actually only empty schemata; and in their application, one must think of whole formulas in the places of A and B – perhaps extended equations, congruences, projections. Then the matter appears completely different. The disadvantage of the waste of space of the "conceptual notation" is converted into the advantage of perspicuity; the advantage of terseness for Boole is transformed into the disadvantage of unintelligibility. The "conceptual notation" makes the most of the two-dimensionality of the writing surface by allowing the assertible contents to follow one below the other while each of these extends [separately] from left to right. Thus, the separate contents are clearly separated from each other, and yet their logical relations are easily visible at a glance. For Boole, a single line, often excessively long, would result. Surely it would be unjust to charge Boole, who never thought of such an application of his formulas, with the easily discoverable disadvantages that arise in this way. But it would be equally unfair to consider as a defect of the "conceptual notation" the waste of space in the case of the mere indication of content [as in the above case where A and B merely indicate extensive complex contents].

Another comment of Schröder is related to what I have just said – my formula language indulges in the Japanese custom of writing vertically. Actually, this appears to be so, as long as one presents only the abstract logical forms. But if one imagines replacing the single letters with whole formulas, say arithmetical equations, he discovers that nothing unusual is presented here; for in every arithmetical derivation, one does not write the separate equations next to each other, but puts them one below the other for the sake of perspicuity.

Thus, in his evaluation, Schröder proceeds throughout from [the assumption of] an immediate comparability – which is non-existent – of the "conceptual notation" and the Leibnizian-Boolean formula language. He means to contribute most effectively to the correction of opinions with the remark that the two modes of symbolization are not essentially different because we could translate from one into the other. But this proves nothing. If the same department of knowledge is symbolized by means of two symbol systems, then it follows necessarily that a translation or transcription from one into the other would be possible. Conversely, nothing more follows from this possibility than the existence of a common department of knowledge. The system of symbols, nevertheless, can be essentially different. We can ask whether this translation is feasible throughout, or whether perhaps my formal language governs a smaller region.*

Schröder says that my "conceptual notation" has almost nothing in common with the Boolean calculus of concepts. From this account, it could appear that the "conceptual notation" would not be able to represent the subordination of concepts. An example will convince one of the opposite. The judgement

$$\vdash \begin{array}{l} x^4 = 81 \\ x^2 = 9 \end{array}$$

runs in words: if $x^2 = 9$, then $x^4 = 81$. Now we can call a number whose square is 9 "a square root of 9", and one whose fourth power is 81 "a fourth root of 81", and then translate: all square roots of 9 are fourth roots of 81. Here the concept "square root of 9" is subordinated to the concept "fourth root of 81". The purpose of the letter x is to make the whole judgement general in the sense that the content should hold whatever one may put in for x. A correct judgement even results if, for example, we put 1 in for x:

$$\vdash \begin{array}{l} 1^4 = 81 \\ 1^2 = 9 \end{array}$$

for the case where $1^2 = 9$ and 1^4 does not equal 81 is to be denied since 1^2 does not equal 9.

It is sometimes necessary to confine the generality to a part of the judgement. Then I make use of German instead of italic letters, as in

$$\vdash \mathfrak{a} \begin{array}{l} x = 0 \\ \mathfrak{a} = x \\ \mathfrak{a}^2 = x \end{array}$$

---

* In this paragraph Frege seems to be giving too much credit to Boolean logic, for *that* is the one which "governs a smaller region". Because of his functional calculus and quantification Frege can express statements and logical relations that Boolean logic cannot handle. Perhaps Frege did not yet (1882) realize what a great advance he had achieved in logic [Translator's note.].

in words: if each square root of x is x itself, then $x = 0$. Here the concavity with the ɑ signifies that the generality expressed by ɑ should be confined to the content of

$$\begin{array}{c} \vphantom{} \\ \end{array} \begin{array}{l} \text{ɑ} = x \\ \text{ɑ}^2 = x. \end{array}$$

I consider this mode of notation one of the most important components of my "conceptual notation", through which it also has, as a mere presentation of logical forms, a considerable advantage over Boole's mode of notation. In this way, in place of the artificial Boolean elaboration, an organic relation between the *primary* and the *secondary propositions* is established. Schröder recognizes the advantage in this, when he makes the effort to establish it in the Boolean formula language. However, in so doing, he shows that he has not grasped the heart of the matter; namely, the delimitation of the scope to which the generality should extend. According to Schröder's proposal, the difference between

$$\begin{array}{l} x = 0 \\ \text{ɑ} = x \\ \text{ɑ}^2 = x \end{array} \quad \text{and} \quad \begin{array}{l} x = 0 \\ \text{ɑ} = x \\ \text{ɑ}^2 = x \end{array}$$

could not be clearly perceived. And yet the difference is so great that the latter is false while the former is true. A further disadvantage of his proposal is that it requires yet another symbol for negation.

It would lead too far afield if I were to answer all of Schröder's particular criticisms. It may suffice to have his false conception of the aim of the "conceptual notation" corrected and to show in this way the lack of cogency of at least some of his critical remarks. Had he attempted to translate some of the formulas of the third chapter of my book and those which I had the honour of presenting to you some time ago into the mode of notation which he calls better, then he would have discovered by the difficulty of the undertaking the erroneous nature of his view.

Nevertheless, I am thankful to him for the review of my work.

4
THE FOUNDATIONS OF ARITHMETIC

Introduction

When we ask someone what the number one is, or what the symbol 1 means, we get as a rule the casual answer "a thing". And if we go on to point out that the proposition

"the number one is a thing"

is not a definition, because it has the definite article on one side and the indefinite on the other, or that it only assigns the number one to the class of things, without stating which thing it is, then we shall very likely be invited to select something for ourselves – anything we please – to call one. Yet if everyone had the right to understand by this name whatever he pleased, then the same proposition about one would mean different things for different people, – such propositions would have no common content. Some will, perhaps, decline to answer the question, pointing out that it is impossible to state, either, what is meant by the letter a, as it is used in arithmetic; and that if we try saying "a means a number," this is open to precisely the same objection as the definition "one is a thing." Now in the case of a it is quite right to decline to answer: a does not mean some one definite number which can be specified, but serves to express the generality of general propositions. If, in $a + a - a = a$, we put for a some number, any we please but the same throughout, we get always a true identity. This is the sense in which the letter a is used. With one, however, the position is essentially different. Can we, in the identity of $1 + 1 = 2$, put for 1 in both places some one and the same object, say the Moon? On the contrary, it looks as though, whatever we put for the first 1, we must put something different for the second. Why is it that we have to do here precisely what would have been wrong in the other case? Again, arithmetic cannot get along with a alone, but has to use further letters besides (b, c and so on), in order to express in general form relations between different numbers. It would therefore be natural to suppose that the symbol 1 too, if it served in some similar way to confer generality of propositions, could not be enough by itself. Yet surely the number one looks like a definite particular object, with properties that can be specified, for example, that of remaining unchanged when multiplied by itself? In this sense, a has no properties that can be specified, since whatever can be asserted of a is a common property of all numbers, whereas $1^1 = 1$ asserts nothing of the Moon, nothing of the Sun, nothing of the Sahara, nothing of the Peak of Teneriffe; for what could be the sense of any such assertion?

Questions like these catch even mathematicians for that matter, or most of them, unprepared with any satisfactory answer. Yet is it not a scandal that our science should be so unclear about the first and foremost among its objects, and one which is apparently so simple? Small hope, then, that we shall be able to say what number is. The fact is, surely, that if a concept fundamental to a mighty science gives rise to difficulties, it is an imperative task to investigate it more closely until those difficulties are overcome; especially as we shall hardly succeed in finally clearing up negative numbers, or fractional or complex numbers, so long as our insight into the foundation of the whole structure of arithmetic is still defective.

Many people will be sure to think this is not worth the trouble. Naturally, they suppose, this concept is adequately dealt with in the elementary textbooks, where the subject is settled and closed for the rest of one's natural lifetime. Who can believe that he has anything still to learn on so simple a matter? So free from all difficulty is the concept of positive whole number held to be, that an account of it fit for children can be both scientific and exhaustive; and that every schoolboy, without any further reflexion or acquaintance with what others have thought, knows all there is to know about it. The first prerequisite for learning anything is thus utterly lacking – I mean, the knowledge that we do not know. The result is that we still rest content with the crudest of views, even though since HERBART's[1] day a better doctrine has been available. It is wearisome and depressing to observe how discoveries once made are always threatening to be lost again in this way, and how much work promises to have been done in vain, because the citizens of the Republic of Letters think it unnecessary to assimilate its results. My work too, as I am well aware, is exposed to this risk. Typical of what I am faced with, typically crude, is the description of calculation as "combinative mechanical thought".[2] I doubt whether there exists any thought whatsoever answering to this description. A combinative imagination, even, might sooner be let pass; but that has no relevance to calculation. Thought is in essentials the same everywhere: it is not true that there are different kinds of laws of thought to suit the different kinds of objects thought about. Such differences as there are consist only in this, that the thought is more pure or less pure, less dependent or more upon psychological influences and on external aids such as words or numerals, and further to some extent too in the finer or coarser structure of the concepts involved; but it is precisely in this respect that mathematics claims to surpass all other sciences, even philosophy.

---

1 Collected Works, ed. Hartenstien, Vol. X, part i, *Umriss pädagogischer Vorlesungen*, § 252, n. 2: "Two does not mean two things, but doubling" etc.
2 K. Fischer, *System der Logik und Metaphysik oder Wissenschaftslebre*, 2nd edn., § 94.

The present work will serve to show that even inferences which on the face of it are peculiar to mathematics, such as that from n to $n + 1$, are based on the general laws of logic, and that there is no need of special laws for combinative thought. It is possible, of course, to operate with figures mechanically, just as it is possible to speak like a parrot: but that hardly deserves the name of thought. It only becomes possible at all after the mathematical notation has, as a result of genuine thought, been so developed that it does the thinking for us, so to speak. This does not prove that numbers are formed in some peculiarly mechanical way, as sand, say, is formed out of quartz granules. In their own interests mathematicians should, I consider, combat any view of this kind, since it is calculated to lead to the disparagement of a principal object of their study, and of their science itself along with it. Yet even in the works of mathematicians are to be found expressions of exactly the same sort. The truth is the very opposite: the concept of number, as we shall be forced to recognize, has a finer structure than most of the concepts of the other sciences, even although it is still one of the simplest in arithmetic.

In order, then, to dispel this illusion that the positive whole numbers really present no difficulties at all, but that universal concord reigns about them, I have adopted the plan of criticizing some of the views put forward by mathematicians and philosophers on the questions raised. It will be seen how small is the extent of their agreement – so small, that precisely contrary opinions are expressed. For example, some hold that "units are identical with one another," others that they are different, and each side supports its assertion with arguments that cannot be rejected out of hand. My object in this is to awaken a desire for a stricter enquiry. At the same time this preliminary examination of the views others have put forward should clear the ground for my own account, by convincing my readers in advance that these other paths do not lead to the goal, and that my opinion is not just one among many all equally tenable; and in this way I hope to settle the question finally, at least in essentials.

I realize that, as a result, my arguments will be of a more philosophical cast than many mathematicians may approve; but any thorough investigation of the concept of number is bound always to turn out rather philosophical. It is a task which is common to mathematics and philosophy.

It may well be that the co-operation between these two sciences, in spite of many démarches from both sides, is not so flourishing as could be wished and would, for that matter, be possible. And if so, this is due in my opinion to the preponderance in philosophy of psychological methods of argument, which have penetrated even into the field of logic. With this tendency mathematics is completely out of sympathy, and this easily accounts for the aversion to philosophical arguments felt by many mathematicians. When STRICKER,[1] for instance, calls our ideas of numbers motor phenomena and makes them dependent on muscular sensations,

---

[1] *Studien über Association der Vorstellungen*, Vienna 1883.

no mathematician can recognize his numbers in such stuff or knows where to begin to tackle such a proposition. An arithmetic founded on muscular sensations would certainly turn out sensational enough, but also every bit as vague as its foundation. No, sensations are absolutely no concern of arithmetic. No more are mental pictures, formed from the amalgamated traces of earlier sense-impressions. All these phases of consciousness are characteristically fluctuating and indefinite, in strong contrast to the definiteness and fixity of the concepts and objects of mathematics. It may, of course, serve some purposes to investigate the ideas and changes of ideas which occur during the course of mathematical thinking; but psychology should not imagine that it can contribute anything whatever to the foundation of arithmetic. To the mathematician as such these mental pictures, with their origins and their transformations, are immaterial. STRICKER himself states that the only idea he associated with the word "hundred" is the symbol 100. Others may have the idea of the letter C or something else; does it not follow, therefore, that these mental pictures are, so far as concerns us and the essentials of our problem, completely immaterial and accidental – just as accidental as it is that a blackboard and a piece of chalk do not by any means qualify to be called ideas of the number a hundred? Then let us cease to suppose that the essence of the matter lies in such ideas. Never again let us take a description of the origin of an idea for a definition, or an account of the mental and physical conditions on which we become conscious of a proposition for a proof of it. A proposition may be thought, and again it may be true; never again let us confuse these two things. We must remind ourselves, it seems, that a proposition no more ceases to be true when I cease to think of it than the sun ceases to exist when I shut my eyes. Otherwise, in proving Pythagoras' theorem we should be reduced to allowing for the phosphorous content of the human brain; and astronomers would hesitate to draw any conclusions about the distant past, for fear of being charged with anachronism, – with reckoning twice two as four regardless of the fact that our idea of number is a product of evolution and has a history behind it. It might be doubted whether by that time it had progressed so far. How could they profess to know that the proposition $2 \times 2 = 4$ was already in existence in that remote epoch? Might not the creatures then extant have held the proposition $2 \times 2 = 5$, from which the proposition $2 \times 2 = 4$ was only evolved later through a process of natural selection in the struggle for existence? Why, it might even be that $2 \times 2 = 4$ itself is destined in the same way to develop into $2 \times 2 = 3$! *Est modus in rebus, sunt certi denique fines*! The historical approach, with its aim of detecting how things begin and of arriving from these origins at a knowledge of their nature, is certainly perfectly legitimate; but it has also its limitations. If everything were in continual flux, and nothing maintained itself fixed for all time, there would no longer be any possibility of getting to know anything about the world and everything would be plunged in confusion. We suppose, it would seem, that concepts grow in the individual mind like leaves on a tree, and we think to discover their nature by studying their growth: we seek to define them psychologically, in terms of the human mind. But this account makes everything subjective, and if we follow it through to the end, does away with truth. What is known as the history of concepts is really a history either of our knowledge of concepts or of the meaning of words. Often it is only after immense intellectual effort, which may have continued over centuries, that humanity at last succeeds in achieving knowledge of a concept in its pure form, in stripping off the irrelevant accretions which veil it from the eyes of the mind. What, then, are we to say of those who, instead of advancing this work where it is not yet completed, despise it, and betake themselves to the nursery, or bury themselves in the remotest conceivable periods of human evolution, there to discover, like JOHN STUART MILL, some gingerbread or pebble arithmetic! It remains only to ascribe to the flavour of the bread some special meaning for the concept of number. A procedure like this is surely the very reverse of rational, and as unmathematical, at any rate, as it could well be. No wonder the mathematicians turn their backs on it. Do the concepts, as we approach their supposed sources, reveal themselves in peculiar purity? Not at all; we see everything as through a fog, blurred and undifferentiated. It is as though everyone who wished to know about America were to try to put himself back in the position of Columbus, at the time when he caught the first dubious glimpse of his supposed India. Of course, a comparison like this proves nothing; but it should, I hope, make my point clear. It may well be that in many cases the history of earlier discoveries is a useful study, as a preparation for further researches; but it should not set up to usurp their place.

So far as mathematicians are concerned, an attack on such views would indeed scarcely have been necessary; but since I had also the philosophers in mind, in trying to settle the issues raised, wherever I could, in a way to convince them too, I found myself forced to enter a little into psychology, if only to repel its invasion of mathematics.

Besides, even mathematical textbooks do at times lapse into psychology. When the author feels himself obliged to give a definition, yet cannot, then he tends to give at least a description of the way in which we arrive at the object or concept concerned. These cases can easily be recognized by the fact that such explanations are never referred to again in the course of the subsequent exposition. For teaching purposes, introductory devices are certainly quite legitimate; only they should always be clearly distinguished from definitions. A delightful example of the way in which even mathematicians can confuse the grounds of proof with the mental or physical conditions to be satisfied if the proof is to be given is to be found in E. SCHRÖDER.[1] Under the heading "Special Axiom" he produces the following: "The principle I have in mind might well be called the Axiom of symbolic Stability. It guarantees us that throughout all our arguments and deductions the symbols remain constant in our memory – or preferably on paper," and so on.

---

1 *Lehrbuch der Arithmetik und Algebra*, [Leipzig 1873].

No less essential for mathematics than the refusal of all assistance from the direction of psychology, is the recognition of its close connexion with logic. I go so far as to agree with those who hold that it is impossible to effect any sharp separation of the two. This much everyone would allow, that any enquiry into the cogency of a proof or the justification of a definition must be a matter of logic. But such enquiries simply cannot be eliminated from mathematics, for it is only through answering them that we can attain to the necessary certainty.

In this direction too I go, certainly further than is usual. Most mathematicians rest content, in enquiries of this kind, when they have satisfied their immediate needs. If a definition shows itself tractable when used in proofs, if no contradictions are anywhere encountered, and if connexions are revealed between matters apparently remote from one another, this leading to improvements in order and increased regularity, it is usual to regard the definition as sufficiently established, and few questions are asked as to its logical justification. This procedure has at least the advantage that it makes it difficult to miss the mark altogether. Even I agree that definitions must show their worth by their fruitfulness: it must be possible to use them for constructing proofs. Yet it must still be borne in mind that the rigour of the proof remains an illusion, however flawless the chain of deductions, so long as the definitions are justified only as an afterthought, by our failing to come across any contradiction. By these methods we shall, at bottom, never have achieved more than an empirical certainty, and we must really face the possibility that we may still in the end encounter a contradiction which brings the whole edifice down in ruins. For this reason I have felt bound to go back rather further into the general logical foundations of our science than perhaps most mathematicians will consider necessary.

In the enquiry that follows, I have kept to three fundamental principles:

- always to separate sharply the psychological from the logical, the subjective from the objective;
- never to ask for the meaning of a word in isolation, but only in the context of a proposition;
- never to lose sight of the distinction between concept and object.

In compliance with the first rule, I have used the word "idea" always in the psychological sense, and have distinguished ideas from concepts and from objects. If the second rule is not observed, one is almost forced to take as the meanings of words mental pictures or acts of the individual mind, and so to offend against the first rule as well. As to the third point, it is a mere illusion to suppose that a concept can be made an object without altering it. From this it follows that a widely-held formalist theory of fractional, negative, etc., numbers is untenable. How I propose to improve upon it can be no more than indicated in the present work. With numbers of all these types, as with the positive whole numbers, everything turns on fixing the sense of an identity.

My results will, I think, at least in essentials, win the adherence of those mathematicians who take the trouble to attend to my arguments. They seem to me to be already in the air, and it may be that every one of them singly, or at least something very like it, has been previously put forward; though perhaps, presented as they are here in connexion with each other, they may still be novel. I have often been astonished at the way in which writers who on one point approach my view so closely, on others depart from it so violently.

Their reception by philosophers will be varied, depending on each philosopher's own position; but presumably those empiricists who recognize induction as the sole original process of inference (and even that as a process not actually of inference but of habituation) will like them least. Here and there, perhaps, some stray philosopher will grasp the opportunity of putting the principles of his theory of knowledge to the test afresh. To those who feel inclined to criticize my definitions as unnatural, I would suggest that the point here is not whether they are natural, but whether they go to the root of the matter and are logically beyond criticism.

I permit myself the hope that even the philosophers, if they examine what I have written without prejudice, will find in it something of use to them.

5
FUNCTION AND CONCEPT

Rather a long time ago* I had the honour of addressing this Society about the symbolic system that I entitled *Conceptual Notation*. To-day I should like to throw light upon the subject from another side, and tell you about some supplementations and new conceptions, whose necessity has occurred to me since then. There can here be no question of setting forth my ideology in its entirety, but only of elucidating some fundamental ideas.

My starting-point is what is called a function in mathematics. The original reference of this word was not so wide as that which it has since obtained; it will be well to begin by dealing with this first usage, and only then consider the later extensions. I shall for the moment be speaking only of functions of a single argument. The first place where a scientific expression appears with a clear-cut reference is where it is required for the statement of a law. This case arose, as regards the function, upon the discovery of higher Analysis. Here for the first time it was a matter of setting forth laws holding for functions in general. So we must go back to the time when higher Analysis was discovered, if we want to know what the word 'function' was originally taken to mean. The answer that we are likely to get to this question is: 'A function of x was taken to be a mathematical expression containing x, a formula containing the letter x.'

Thus, e.g., the expression

$$2x^3 + x$$

would be a function of x, and

$$2.2^3 + 2$$

would be a function of 2. This answer cannot satisfy us, for here no distinction is made between form and content, sign and thing signified; a mistake, admittedly, that is very often met with in mathematical works, even those of

---

* On January 10, 1879, and January 27, 1882

celebrated authors. I have already pointed out on a previous occasion* the defects of the current formal theories in arithmetic. We there have talk about signs that neither have nor are meant to have any content, but nevertheless properties are ascribed to them which are unintelligible except as belonging to the content of a sign. So also here; a mere expression, the form for a content, cannot be the heart of the matter; only the content itself can be that. Now what is the content of '$2.2^3 + 5$'? What does it stand for? The same thing as '18' or '3.6.' What is expressed in the equation '$2.2^3 + 2 = 18$' is that the right-hand complex of signs has the same reference as the left-hand one, I must here combat the view that, e.g., 2 + 5 and 3 + 4 are equal but not the same. This view is grounded in the same confusion of form and content, sign and thing signified. It is as though one wanted to regard the sweet-smelling violet as differing from *Viola odorata* because the name sounds different. Difference of sign cannot by itself be a sufficient ground for difference of the thing signified. The only reason why in our case the matter is less obvious is that the reference of the numeral 7 is not anything perceptible to the senses. There is at present a very widespread tendency not to recognize as an object anything that cannot be perceived by means of the senses; this leads here to numerals' being taken to be numbers, the proper objects of our discussions;[†] and then, I admit 7 and 2 + 5 would indeed be different. But such a conception is untenable, for we cannot speak of any arithmetical properties of numbers whatsoever without going back to what the signs stand for. For example, the property belonging to 1, of being the result of multiplying itself by itself, would be a mere myth; for no microscopical or chemical investigation, however far it was carried, could ever detect this property in the possession of the innocent character that we call a figure one. Perhaps there is talk of a definition; but no definition is creative in the sense of being able to endow a thing with properties that it has not already got — apart from the one property of expressing and signifying something in virtue of the definition.[‡] The characters we call numerals have, on the other hand, physical and chemical properties depending on the writing material. One could imagine the introduction some day of quite new numerals, just as, e.g., the Arabic numerals superseded the Roman. Nobody is seriously going to suppose that in this way we should get quite new numbers, quite new arithmetical objects, with properties still to be investigated. Thus we must distinguish between numerals and what they stand for; and if so, we shall have to recognize that the expression '2,' '1 + 1,' '3 – 1,' '6:3' stand for the same thing, for it is quite inconceivable where the difference between them could lie. Perhaps you say: 1 + 1 is a sum, but 6:3 is

---

* *Foundations of Arithmetic*, Breslau, 1884.
† Cf. the essays: *Zählen und Messen erkenntnistheoretisch betrachtet*, by H. von Hemholtz, and *Ueber den Zahlbegriff*, by Leopold Kronecker (*Philosophische Aufsätze. Eduard Zeller zu seinen fünfzigjährigen Doctorjubiläum gewidmet*, Leipzig, 1687).
‡ In definition it is always a matter of associating with a sign a sense or a reference. Where sense and reference are missing, we cannot properly speak either of a sign or of a definition.

a quotient. But what is 6:3? The number that when multiplied by 3 gives the result 6. We say '*the* number,' not 'a number'; by using the definite article, we indicate that there is only a single number. Now we have:

$$(1 + 1) + (1 + 1) + (1 + 1) = 6,$$

and thus $(1 + 1)$ is the very number that was designated as $(6:3)$. The different expressions correspond to different conceptions and aspects, but nevertheless always to the same thing. Otherwise the equation $x^2 = 4$ would not have just the roots 2 and –2, but also the root $(1 + 1)$ and countless others, all of them different, even if they resembled one another in a certain respect. By recognizing only two real roots, we are rejecting the view that the sign of equality does not stand for complete coincidence but only for partial agreement. If we adhere to this truth, we see that the expressions:

$$'2.1^3 + 1,'$$
$$'2.2^3 + 2,'$$
$$'2.4^3 + 4,'$$

stand for numbers, viz. 3, 18, 132. So if the function were really the reference of a mathematical expression, it would just be a number; and nothing new would have been gained for arithmetic [by speaking of functions]. Admittedly, people who use the word 'function' ordinarily have in mind expressions in which a number is just indicated indefinitely by the letter x, e.g.

$$'2.x^3 + x';$$

but that makes no difference; for this expression likewise just indicates a number indefinitely, and it makes no essential difference whether I write it down or just write down 'x.'

All the same, it is precisely by the notation that uses 'x' to indicate [a number] indefinitely that we are led to the right conception. People call x the argument, and recognize the same function again in

$$'2.1^3 + 1,'$$
$$'2.4^3 + 4,'$$
$$'2.5^3 + 5,'$$

only with different arguments, viz. 1, 4, and 5. From this we may discern that it is the common element of these expressions that contains the essential peculiarity of a function; i.e. what is present in

$$'2.x^3 + x'$$

over and above the letter 'x.' We could write this somewhat as follows:

$$\text{'}2.(\ )^3 + (\ ).\text{'}$$

I am concerned to show that the argument does not belong with the function, but goes together with the function to make up a complete whole; for the function by itself must be called incomplete, in need of supplementation, or 'unsaturated.' And in this respect functions differ fundamentally from numbers. Since such is the essence of the function, we can explain why on the one hand, we recognize the same function in '$2.1^3 + 1$' and '$4 - 1$' in spite of their equal numerical values. Moreover, we now see how people are easily led to regard the form of the expression as what is essential to the function. We recognize the function in the expression by imagining the latter as split up, and the possibility of thus splitting it up is suggested by its structure.

The two parts into which the mathematical expression is thus split up, the sign of the argument and the expression of the function, are dissimilar; for the argument is a number, a whole complete in itself, as the function is not. (We may compare this with the division of a line by a point. One is inclined in that case to count the dividing-point along with both segments; but if we want to make a clean division, i.e. so as not to count anything twice over or leave anything out, then we may only count the dividing-point along with one segment. This segment thus becomes fully complete in itself, and may be compared to the argument; whereas the other is lacking in something — viz. the dividing-point, which one may call the endpoint, does not belong to it. Only by completing it with this endpoint, or with a line that has two endpoints, do we get from it something entire.) For instance, if I say 'the function $2.x^3 + x$,' x must not be considered as belonging to the function; this letter only serves to indicate the kind of supplementation that is needed; it enables one to recognize the places where the sign for the argument must go in.

We give the name 'the value of a function for an argument' to the result of completing the function with the argument. Thus, e.g., 3 is the value of the function $2.x^3 + 1 = 3$.

There are functions, such as $2 + x - x$ or $2 + 0.x$, whose value is always the same, whatever the argument; we have $2 = 2 + x - x$ and $2 = 2 + 0.x$. Now if we counted the argument as belonging with the function, we should hold that the number 2 is this function. But this is wrong. Even though here the value of the function is always 2, the function itself must nevertheless be distinguished from 2; for the expression for a function must always show one or more places that are intended to be filled up with the sign of the argument.

The method of analytic geometry supplies us with a means of intuitively representing the values of a function for different arguments. If we regard the argument as the numerical value of an abscissa, and the corresponding value of the function as the numerical value of the ordinate of a point, we obtain a set of points that presents itself to intuition (in ordinary cases) as a curve. Any point on the curve corresponds to an argument together with the associated value of the function. Thus, e.g.,

$$y = x^2 - 4x$$

yields a parabola; here 'y' indicates the value of the function and the numerical value of the ordinate, and 'x' similarly indicates the argument and the numerical value of the abscissa. If we compare with this the function

$$x(x - 4),$$

we find that they have always the same value for the same argument. We have generally:

$$x^2 - 4x = x(x - 4),$$

whatever number we take for x. Thus the curve we get from

$$y = x^2 - 4x$$

is the same as the one that arises out of

$$y = x(x - 4).$$

I express this as follows: the function $x(x - 4)$ has the same range of values as the function $x^2 - 4x$.

If we write

$$x^2 - 4x = x(x - 4),$$

we have not put one function equal to the other, but only the values of one equal to those of the other. And if we so understand this equation that it is to hold whatever argument may be substituted for x, then we have thus expressed that an equality holds generally. But we can also say: 'the value-range of the function $x(x - 4)$ is equal to that of the function $x^2 - 4x$,' and here we have an equality between ranges of values. The possibility of regarding the equality holding generally between values of functions as a [particular] equality, viz. an equality between ranges of values, is, I think, indemonstrable; it must be taken to be a fundamental law of logic.*

---

* In many phrases of ordinary mathematical terminology, the word 'function' certainly corresponds to what I have here called the value-range of a function. But function, in the sense of the word employed here, is the logically prior [notion].

We may further introduce a brief notation for the value-range of a function. To this end I replace the sign of the argument in the expression for the function by a Greek vowel, enclose the whole in brackets, and prefix to it the same Greek letter with a smooth breathing. Accordingly, e.g.,

$$\grave{\epsilon}(\epsilon^2 - 4\epsilon)$$

is the value-range of the function $x^2 - 4x$ and

$$\grave{\alpha}(\alpha.(\alpha - 4))$$

is the value-range of the function $x(x - 4)$, so that in

$$`\grave{\epsilon}(\epsilon^2 - 4\epsilon) = \grave{\alpha}(\alpha.(\alpha - 4))\text{'}$$

we have the expression for: the first range of values is the same as the second. A different choice of Greek letters is made on purpose, in order to indicate that there is nothing that obliges us to take the same one.

If we understand

$$`x^2 - 4x = x(x - 4)\text{'}$$

in the same sense as before, this expresses the same sense, but in a different way. It presents the sense as an equality holding generally; whereas the newly-introduced expression is simply an equation, and its right side, and equally its left side, stands for something complete in itself. In

$$`x^2 - 4x = x(x - 4)\text{'}$$

the left side considered in isolation indicates a number only indefinitely, and the same is true of the right side. If we just had '$x^2 - 4x$' we could write instead '$y^2 - 4y$' without altering the sense; for 'y' like 'x' indicates a number only indefinitely. But if we combine the two sides to form an equation, we must choose the same letter for both sides, and we thus express something that is not contained in the left side by itself, nor in the right side, nor in the 'equals' sign; viz. generality. Admittedly what we express is the generality of an equality; but primarily it is a generality.

Just as we indicate a number indefinitely by a letter, in order to express generality, we also need letters to indicate a function indefinitely. To this end people ordinarily use the letters f and F, thus '$f(x)$,' '$F(x)$,' where 'x' replaces the argument. Here the need of the function for supplementation is expressed by the fact that the letter f or F carries along with it a pair of brackets; the space between these is meant to receive the sign for the argument. Thus

$$\acute{\epsilon}\, F(\epsilon)$$

indicates the value-range of a function that is left undetermined.

Now how has the reference of the word 'function' been extended by the progress of science? We can distinguish two directions in which this has happened.

In the first place, the field of mathematical operations that serve for constructing functions has been extended. Besides addition, multiplication, exponentiation, and their converses, the various means of transition to the limit have been introduced – to be sure, without people's being always clearly aware that they were thus adopting something essentially new. People have gone further still, and have actually been obliged to resort to ordinary language, because the symbolic language of Analysis failed; e.g. when they were speaking of a function whose value is 1 for rational and 0 for irrational arguments.

Secondly, the field of possible arguments and values for functions has been extended by the admission of complex numbers. In conjunction with this, the sense of the expressions 'sum,' 'product,' etc., had to be defined more widely.

In both directions I go still further. I begin by adding to the signs +, –, etc., which serve for constructing a functional expression, also signs such as =, >, <, so that I can speak, e.g., of the function $x^2 = 1$, where x takes the place of the argument as before. The first question that arises here is what the values of this function are for different arguments. Now if we replace x successively by –1, 0, 1, 2, we get:

$$(-1)^2 = 1,$$
$$0^2 = 1,$$
$$1^2 = 1,$$
$$2^2 = 1.$$

Of these equations the first and third are true, the others false. I now say: 'the value of our function is a truth-value' and distinguish between the truth-values of what is true and what is false. I call the first, for short, the True; and the second, the False. Consequently, e.g., '$2^2 = 4$' stands for the True as, say, '2^2' stands for 4. And '$2^2 = 1$' stands for the False. Accordingly

'$2^2 = 4$,' '$2 > 1$,' '$2^4 = 4^2$,'

stand for the same thing, viz. the True, so that in

$$(2^2 = 4) = (2 > 1)$$

we have a correct equation.

The objection here suggests itself that '$2^2 = 4$' and '$2 > 1$' nevertheless make quite different assertions, express quite different thoughts; but likewise '$2^4 = 4^2$' and '$4.4 = 4^2$,' express different thoughts; and yet we can replace '2^4' by '4.4,' since both signs have the same reference. Consequently, '$2^4 = 4^2$' and '$4.4 = 4^2$' likewise have the same reference. We see from this that from identity of reference there does not follow identity of the thought [expressed]. If we say 'the Evening Star is a planet with a shorter period of revolution than the Earth,' the thought we express is other than in the sentence 'the Morning Star is a planet with a shorter period of revolution than the Earth'; for somebody who does know that the Morning Star is the Evening Star might regard one as true and the other as false. And yet both sentences must have the same reference; for it is just a matter of interchanging the words 'Evening Star' and 'Morning Star,' which have the same reference, i.e. are proper names of the same heavenly body. We must distinguish between sense and reference. '2^4' and '4^2' certain have the same reference, i.e. they are proper names of the same number; but they do not have the same sense; consequently '$2^4 = 4^2$' and '$4.4 = 4^2$,' have the same reference, but not the same sense (which means, in this case: they do not contain the same thought).*

Thus, just as we write:

$$'2^4 = 4.4'$$

we may also write with equal justification

$$'(2^4 = 4^2) = (4.4 = 4^2)'$$
$$\text{and} \quad '(2^2 = 4) = (2 > 1).'$$

It might further be asked: What, then, is the point of admitting the signs =, >, <, into the field of those that help to build up a functional expression? Nowadays, it seems, more and more supporters are being won by the view that arithmetic is a further development of logic; that a more rigorous establishment of arithmetical laws reduces them to purely logical laws and to such laws alone. I too am of this opinion, and I base upon it the requirement that the symbolic language of arithmetic must be expanded into a logical symbolism. I shall now have to indicate how this is done in our present case.

We saw that the value of our function $x^2 = 1$ is always one of the two truth-values. Now if for a definite argument, e.g. – 1, the value of the function is the True, we can express this as follows: 'the number – 1 has the property that its square is 1'; or, more briefly, '– 1 is a square root of 1'; or, '– 1 falls under the concept: square root of 1.' If the value of the function $x^2 = 1$ for an argument, e.g. for 2, is the False, we can express this as follows: '2 is not a square root of 1' or '2 does not fall under the concept: square root of 1.' We thus see how closely that which is

---

* I do not fail to see that this way of putting it may at first seem arbitrary and artificial, and that it would be desirable to establish my view by going further into the matter. Cf. my forthcoming essay ['Sense and Reference'].

called a concept in logic is connected with what we call a function. Indeed, we may say at once: a concept is a function whose value is always a truth-value. Again, the value of the function

$$(x + 1)^2 = 2(x + 1)$$

is always a truth-value. We get the True as its value, e.g., for the argument -1, and this can also be expressed thus: -1 is a number less by 1 than a number whose square is equal to its double. This expresses the fact that -1 falls under a concept. Now the functions

$$x^2 = 1 \text{ and } (x + 1)^2 = 2(x + 1)$$

always have the same value for the same argument, viz. the True for the arguments -1 and $+1$, and the False for all other arguments. According to our previous conventions we shall also say that these functions have the same range of values, and express this in symbols as follows:

$$\acute{\epsilon}(\epsilon^2 = 1) = \acute{\alpha}((\alpha + 1)^2 = 2(\alpha + 1))$$

In logic this is called identity of the extension of the concepts. Hence we can designate as an extension the value-range of a function whose value for every argument is a truth-value.

We shall not stop at equations and inequalities. The linguistic form of equations is a statement. A statement contains (or at least purports to contain) a thought as its sense; and this thought is in general true or false; i.e. it has in general a truth-value, which must be regarded as the reference of the sentence, just as (say) the number 4 is the reference of the expression '2 + 2,' or London of the expression 'the capital of England.'

Statements in general, just like equations or inequalities or expressions in Analysis, can be imagined to be split up into two parts; one complete in itself, and the other in need of supplementation, or 'unsaturated.' Thus, e.g. we split up the sentence

'Caesar conquered Gaul'

into 'Caesar' and 'conquered Gaul.' The second part is 'unsaturated' – it contains an empty place; only when this place is filled up with a proper name, or with an expression that replaces a proper name, does a complete sense appear. Here too I give the name 'function' to what this 'unsaturated' part stands for. In this case the argument is Caesar.

We see that here we have undertaken to extend [the application of the term] in the other direction, viz. as regards what can occur as an argument. Not merely numbers, but objects in general, are now admissible; and here persons must assuredly be counted as objects. The two truth-values have already been introduced as possible values of a function; we must go further and admit objects without restriction as values of functions. To get an example of this, let us start, e.g., with the expression

'the capital of the German Empire.'

This obviously takes the place of a proper name, and stands for an object. If we now split it up into the parts

'the capital of' and 'the German Empire'

where I count the [German] genitive form as going with the first part, then this part is 'unsaturated,' whereas the other is complete in itself. So in accordance with what I said before, I call

'the capital of x'

the expression of a function. If we take the German Empire as the argument, we get Berlin as the value of the function.

When we have thus admitted objects without restriction as arguments and values of functions, the question arises what it is that we are here calling an object. I regard a regular definition as impossible, since we have here something too simple to admit of logical analysis. It is only possible to indicate what is meant. Here I can only say briefly: An object is anything that is not a function, so that an expression for it does not contain any empty place.

A statement contains no empty place, and therefore we must regard what it stands for as an object. But what a statement stands for is a truth-value. Thus the two truth-values are objects.

Earlier on we presented equations between ranges of values, e.g.:

$$`\acute{\epsilon}(\epsilon^2 - 4\epsilon) = \acute{\alpha}(\alpha(\alpha - 4)).\text{'}$$

We can split this up into '$\acute{\epsilon}(\epsilon^2 - 4\epsilon)$' and '$(\ ) = \acute{\alpha}(\alpha(\alpha - 4))$.' This latter part needs supplementation, since on the left of the 'equals' sign it contains an empty place. The first part, '$\acute{\epsilon}(\epsilon^2 - 4\epsilon)$,' if fully complete in itself and thus stands for an object. Value-ranges of functions are objects, whereas functions themselves are not. We gave the name 'value-range' also to $\acute{\epsilon}(\epsilon^2 = 1)$, but we could also have termed it the extension of the concept: square root of 1. Extensions of concepts likewise are objects, although concepts themselves are not.

After thus extending the field of things that may be taken as arguments, we must get more exact specifications as to what the signs already in use stand for. So long as the only objects dealt with in arithmetic are the integers, the letters *a* and *b* in '*a* + *b*' indicate only integers, the plus-sign need be defined only between integers. Every extension of the field to which the objects indicated by *a* and *b* belong obliges us to give a new definition of the plus-sign. It seems to be demanded by scientific rigour that we should have provisos against an expression's possibly coming to have no reference; we must see to it that we never perform calculations with empty signs in the belief that we are dealing with objects. People have in the past carried out invalid procedures with divergent infinite series. It is thus necessary to lay down rules from which it follows, e.g., what

$$\odot + 1$$

stands for, if ⊙ is to stand for the Sun. What rules we lay down is a matter of comparative indifference; but it is essential that we should do so – that '*a* + *b*' should always have a reference, whatever signs for definite objects may be inserted in place of '*a*' and '*b*.' This involves the requirement as regards concepts, that, for any argument, they shall have a truth-value as their value; that it shall be determinate, for any object, whether it falls under the concept or not. In other words: as regards concepts we have a requirement of sharp delimitation; if this were not satisfied it would be impossible to set forth logical laws about them. For any argument x for which 'x + 1' were devoid of reference, the function $x + 1 = 10$ would likewise have no value, and thus no truth-value either, so that the concept:

'what gives the result 10 when increased by 1'

would have no sharp boundaries. The requirement of the sharp delimitation of concepts thus carries along with it this requirement for functions in general that they must have a value for every argument.

We have so far considered truth-values only as values of functions, not as arguments. By what I have just said, we must get a value of a function when we take a truth-value as the argument; but as regards the signs already in common use, the only point, in most cases, of a rule to this effect is that there should *be* a rule; it does not much matter what is determined upon. But now we must deal with certain functions that are of importance to use precisely when their argument is a truth-value.

I introduce the following as such a function

$$\qquad —\!\!\!— x$$

I lay down the rule that the value of this function shall be the True if the True is taken as argument, and that contrariwise, in all other cases the value of this function is the False — i.e. both when the argument is the False and when it is not a truth-value at all. Accordingly, e.g.

$$\underline{} 1 + 3 = 4$$

is the True, whereas both

$$\underline{} 1 + 3 = 5$$

and also

$$\underline{} 4$$

are the False. Thus this function has as its value the argument itself, when that is a truth-value. I used to call this horizontal stroke the content-stroke — a name that no longer seems to me appropriate. I now wish to call it simply the horizontal.

If we write down an equation or inequality, e.g. 5 > 4, we ordinarily wish at the same time to express a judgment; in our example, we want to assert that 5 is greater than 4. According to the view I am here presenting, '5 > 4' and '1 + 3 = 5' just give us expressions for truth-values, without making any assertion. This separation of the act from the subject-matter of judgment seems to be indispensable; for otherwise we could not express a mere supposition — the putting of a case without a simultaneous judgment as to its arising or not. We thus need a special sign in order to be able to assert something. To this end I make use of a vertical stroke at the left end of the horizontal, so that, e.g., by writing

$$\vdash \underline{} 2 + 3 = 5$$

we assert that 2 + 3 equals 5. Thus here we are not just writing down a truth-value, as in

$$2 + 3 = 5,$$

but also at the same time saying that it is True.*

The next simplest function, we may say, is the one whose value is the False for just those arguments for which the value of $\underline{}x$ is the True, and, conversely, is the True for the arguments for which the value of $\underline{}x$ is the False. I symbolize it thus:

$$\underline{}_{\top} x,$$

---

* The assertion sign cannot be used to construct a functional expression; for it does not serve, in conjunction with other signs, to designate an object. '$\vdash \underline{} 2 + 3 = 5$' does not designate anything; it asserts something.

Function and Concept   155

and here I call the little vertical stroke the stroke of negation. I conceive of this as a function with the argument —— x:

$$(-\!\!\top\!\!- x) = (-\!\!\top\!\!- (-\!\!\!-\!\!\!- x))$$

where I imagine the two horizontal strokes to be fused together. But we also have:

$$(-\!\!\!-\!\!\!- (-\!\!\top\!\!- x)) = (-\!\!\top\!\!- x),$$

since the value of —⊤— x is always a truth-value. I thus regard the bits of the stroke in —⊤— x to the right and to the left of the stroke of negation as horizontals, in the sense of the word that I defined previously. Accordingly, e.g.:

$$-\!\!\top\!\!- 2^2 = 5$$

stands for the True, and we may add the assertion-sign:

$$\vdash\!\!\top\!\!- 2^2 = 5$$

and in this we assert that $2^2 = 5$ is not the True, or that 2^2 is not 5. But moreover

$$-\!\!\top\!\!- 2$$

is the True, since —— 2 is the False:

$$\vdash\!\!\top\!\!- 2$$

i.e. 2 is not the True.

My way of presenting generality can best be seen in an example. Suppose what we have to express is that every object is equal to itself. In

$$x = x$$

we have a function, whose argument is indicated by 'x.' We now have to say that the value of this function is always the True, whatever we take as argument. I now take the sign

$$-\!\!\overset{a}{\smile}\!\!- f(a)$$

to mean the True when the function $f(x)$ always has the True as its value, whatever the argument may be; in all other cases

$$-\!\!\overset{a}{\smile}\!\!- f(a)$$

is to stand for the False. For our function $x = x$ we get the first case. Thus

$$\overset{a}{\frown} f(a)$$

is the True; and we write this as follows:

$$\vdash \overset{a}{\frown} a = a$$

The horizontal strokes to the right and to the left of the concavity are to be regarded as horizontals in our sense. Instead of 'a,' any other Gothic letter could be chosen; except those which are to serve as letters for a function, like f and F.

This notation affords the possibility of negating generality, as in

$$\underset{\top}{}\overset{a}{\frown} a^2 = 1.$$

That is to say, $\underset{\top}{}\overset{a}{\frown} a^2 = 1$ is the False, since not every argument makes the value of the function $x^2 = 1$ to be the True. (Thus, e.g., we get $2^2 = 1$ for the argument 2, and this is the False.) Now if $\overset{a}{\frown} a^2 = 1$ is the False, then $\underset{\top}{}\overset{a}{\frown} a^2 = 1$ is the True, according to the rule that we laid down previously for the stroke of negation. Thus we have

$$\vdash\underset{\top}{}\overset{a}{\frown} a^2 = 1;$$

i.e. 'not every object is a square root of 1,' or 'there are objects that are not square roots of 1.'

Can we also express: there are square roots of 1? Certainly: we need only take, instead of the function $x^2 = 1$, the function

$$\underset{\top}{} x^2 = 1.$$

By fusing together the horizontals in

$$\overset{a}{\frown} \quad \underset{\top}{} a^2 = 1$$

we get

$$\overset{a}{\frown}\underset{\top}{} a^2 = 1.$$

This stands for the False, since not every argument makes the value of the function

$$\underset{\top}{} x^2 = 1$$

to be the True. E.g.:

$$\vdash 1^2 = 1$$

is the False for $1^2 = 1$ is the True. Now since

$$-\smallfrown^a\!\!\!\!-\vdash a^2 = 1$$

is thus the False,

$$\vdash\!\!-\smallfrown^a\!\!\!\!-\vdash a^2 = 1$$

is the True:

$$\vdash\!\!-\smallfrown^a\!\!\!\!-\vdash a^2 = 1;$$

i.e. 'not every argument makes the value of the function

$$\vdash x^2 = 1$$

to be the True,' or: 'not every argument makes the value of the function $x^2 = 1$ to be the False,' or: 'there is at least one square root of 1.'

At this point there may follow a few examples in symbols and words.

$$\vdash\!\!-\smallfrown^a\!\!\!\!-\vdash a \geqq 0,$$

there is at least one positive number;

$$\vdash\!\!-\smallfrown^a\!\!\!\!-\vdash a < 0,$$

there is at least one negative number;

$$\vdash\!\!-\smallfrown^a\!\!\!\!-\vdash a^3 - 3a^2 + 2a = 0,$$

there is at least one root of the equation

$$x^3 - 3x^2 + 2x = 0.$$

From this we may see how to express existential sentences, which are so important. If we use the functional letter *f* as an indefinite indication of a concept, then

$$-\smallfrown^a\!\!\!\!-\vdash f(a)$$

gives us the form that includes the last examples (if we abstract from the assertion-sign). The expressions

$$\vdash a^2 = 1, \qquad \vdash a \geq 0,$$

$$\vdash a < 0, \qquad \vdash a^3 - 3a^2 + 2a = 0$$

arise from this form in a matter analogous to that in which x^2 gives rise to '1^2,' '2^2,' '3^2.' Now just as in x^2 we have a function whose argument is indicated by 'x,' I also conceive of

$$\vdash f(a)$$

as the expression of a function whose argument is indicated by 'f.' Such a function is obviously a fundamentally different one from those we have dealt with so far; for only a function can occur as its argument. Now just as functions are fundamentally different from objects, so also functions whose arguments are and must be functions are fundamentally different from functions whose arguments are objects and cannot be anything else. I call the latter first-level, the former second-level, functions, In the same way, I distinguish between first-level and second-level concepts.* Second-level functions have actually long been used in Analysis; e.g. definite integrals (if we regard the function to be integrated as the argument).

I will now add something about functions with two arguments. We get the expression for a function by splitting up the complex sign for an object into a 'saturated' and an 'unsaturated' part. Thus, we split up this sign for the True,

$$3 > 2,$$

into '3' and '$x > 2$.' We can further split up the 'unsaturated' part '$x > 2$' in the same way, into '2' and

$$x > y,$$

where 'y' enables us to recognize the empty place previously filled up by '2.' In

$$x > y$$

---

* Cf. my *Foundations of Arithmetic*, Breslau, 1884. I there used the term 'second-order' instead of 'second-level.' The ontological proof of God's existence suffers from the fallacy of treating existence as a first-level concept.

we have a function with two arguments, one indicated by 'x' and the other by 'y'; and in

$$3 > 2$$

we have the value of this function for the arguments 3 and 2. We have here a function whose value is always a truth-value. We called such functions of one argument concepts; we call such functions of two arguments relations. Thus we have relations also, e.g., in

$$x^2 + y^2 = 9$$

and in

$$x^2 + y^2 > 9,$$

whereas the function

$$x^2 + y^2$$

has numbers as values. We shall therefore not call this a relation.

At this point I may introduce a function not peculiar to arithmetic. The value of the function

$$\vdash\!\!\!\!\begin{array}{c} x \\ y \end{array}$$

is to be the False if we take the True as the y-argument and at the same time take some object that is not the True as the x-argument; in all other cases the value of this function is to be the True. The lower horizontal stroke, and the two bits that the upper one is split into by the vertical, are to be regarded as horizontals [in our sense]. Consequently, we can always regard as the arguments of our function ——— x and ——— y, i.e. truth-values.

Among functions of one argument we distinguish first-level and second-level ones. Here, a greater multiplicity is possible. A function of two arguments may be of the same level in relation to them, or of different levels; there are equal-levelled and unequal-levelled functions. Those we have dealt with up to now were equal-levelled. An example of an unequal-levelled function is the differential quotient, if we take the arguments to be the function that is to be differentiated and the argument for which it is differentiated; or the definite integral, so long as we take as arguments the function to be integrated and the upper limit. Equal-levelled functions can again be divided into first-level and second-level ones. An example of a second-level one is

$$F(f(1)),$$

where '*F*' and '*f*' indicate the arguments.

In regard to second-level functions with one argument, we must make a distinction, according as the role of this argument can be played by a function of one or of two arguments; for a function of one argument is essentially so different from one with two arguments that the one function cannot occur as an argument in the same place as the other. Some second-level functions of one argument require that this should be a function with one argument; others, that it should be a function with two arguments; and these two classes are sharply divided.

$$\underline{\quad e \quad d \quad a} \begin{array}{l} d = a \\ F(e,a) \\ F(e,d) \end{array}$$

is an example of a second-level function with one argument, which requires that this should be a function of two arguments. The letter *F* here indicates the argument, and the two places, separated by a comma, within the brackets that follow '*F*' bring it to our notice that *F* represents a function with two arguments.

For functions of two arguments there arises a still greater multiplicity.

If we look back from here over the development of arithmetic, we discern an advance from level to level. At first people did calculations with individual numbers, 1, 3, etc.

$$2 + 3 = 5, \; 2.3 = 6$$

are theorems of this sort. Then they went on to more general laws that hold good for all numbers. What corresponds to this in symbolism is the transition to the literal notation.

A theorem of this sort is

$$(a + b)c = a.c + b.c$$

At this stage they had got to the point of dealing with individual functions; but were not yet using the word, in its mathematical sense, and had not yet formed the conception of what it now stands for. The next higher level was the recognition of general laws about functions, accompanied by the coinage of the technical term 'function.' What corresponds to this in symbolism is the introduction of letters like *f*, *F* to indicate functions indefinitely. A theorem of this sort is

$$\frac{df(x).F(x)}{dx} = F(x).\frac{df(x)}{dx} + f(x).\frac{dF(x)}{dx}$$

Now at this point people had particular second-level functions, but lacked the conception of what we have called second-level functions. By forming that, we make the next step forward. One might think that this would go on. But probably this last step is already not so rich in consequences as the earlier ones; for instead of second-level functions one can deal, in further advances, with first-level functions – as shall be shown elsewhere. But this does not banish from the world the difference between first-level and second-level functions; for it is not made arbitrarily, but founded deep in the nature of things.

Again, instead of functions of two arguments we can deal with functions of a single but complex argument; but the distinction between functions of one and of two arguments still holds in all its sharpness.

6
ON CONCEPT AND OBJECT

In a series of articles in this Quarterly on intuition and its psychical elaboration, Benno Kerry has several times referred to my *Foundations of Arithmetic* and other works of mine, sometimes agreeing and sometimes disagreeing with me. I cannot but be pleased at this, and I think the best way I can show my appreciation is to take up the discussion of the points he contests. This seems to me all the more necessary, because his opposition is at least partly based on a misunderstanding, which might be shared by others, of what I say about the concept; and because, even apart from this special occasion, the matter is important and difficult enough for a more thorough treatment than seemed to me suitable in my *Foundations*.

The word 'concept' is used in various ways; its sense is sometimes psychological, sometimes logical, and sometimes perhaps a confused mixture of both. Since this licence exists, it is natural to restrict it by requiring that when once a usage is adopted it shall be maintained. What I decided was to keep strictly to a purely logical use; the question whether this or that use is more appropriate is one that I should like to leave on one side, as of minor importance. Agreement about the mode of expression will easily be reached when once it is recognized that there is something that deserves a special term.

It seems to me that Kerry's misunderstanding results from his unintentionally confusing his own usage of the word 'concept' with mine. This readily gives rise to contradictions, for which my usage is not to blame.

Kerry contests what he calls my definition of 'concept.' I would remark, in the first place, that my explanation is not meant as a proper definition. One cannot require that everything shall be defined, any more than one can require that a chemist shall decompose every substance. What is simple cannot be decomposed, and what is logically simple cannot have a proper definition. Now something logically simple is no more given us at the outset than most of the chemical elements are; it is reached only by means of scientific work. If something has been discovered that is simple, or at least must count as simple for the time being, we shall have to coin a term for it, since language will not originally contain an expression that exactly answers. On the introduction of a name for something logically simple, a definition is not possible; there is nothing for it but to lead the reader or hearer, by means of hints, to understand the words as is intended.

Kerry wants to make out that the distinction between concept and object is not absolute. 'In a previous passage,' he says, 'I have myself expressed the opinion that the relation between the content of the concept and the concept-object is, in a certain respect, a peculiar and irreducible one; but this was in no way bound up with the view that the properties of being a concept and of being an object are mutually exclusive. The latter view no more follows from the former than it would follow, if, e.g., the relation of father and son were one that could not be further reduced, that a man could not be at once a father and a son (though of course not, e.g., father of the man whose son he was).'

Let us fasten on this simile! If there were, or had been, beings that were fathers but could not be sons, such beings would obviously be quite different in kind from all men who are sons. Now it is something like this that happens here. The concept (as I understand the word) is predicative.* On the other hand, a name of an object, a proper name, is quite incapable of being used as a grammatical predicate. This admittedly needs elucidation, otherwise it might appear false. Surely one can just as well assert of a thing that it is Alexander the Great, or is the number four, or is the planet Venus, as that it is green or is a mammal? If anybody thinks this, he is not distinguishing the usages of the word 'is.' In the last two examples it serves as a copula, as a mere verbal sign of predication. (In this sense [the German word *ist*] can sometimes be replaced by the mere personal suffix: cf. *dies Blatt ist grün* and *dies Blatt grünt*.) We are here saying that something falls under a concept, and the grammatical predicate stands for this concept. In the first three examples, on the other hand, 'is' is used like the 'equals' sign in arithmetic, to express an equation.† In the sentence 'The morning star is Venus,' we have two proper names, 'morning star' and 'Venus,' for the same object. In the sentence 'the morning star is a planet' we have a proper name, 'the morning star,' and a concept-word, 'planet.' So far as language goes, no more has happened than that 'Venus' has been replaced by 'a planet'; bur really the relation has become wholly different. An equation is reversible; an object's falling under a concept is an irreversible relation. In the sentence 'the morning star is Venus,' 'is' is obviously not the mere copula; its content is an essential part of the predicate, so that the word 'Venus' does not constitute the whole of the predicate.‡ One might say instead: 'the morning star is

---

* It is, in fact, the reference of a grammatical predicate.

† I use the word 'equal' and the symbol '=' in the sense 'the same as,' 'no other than,' 'identical with.' Cf. E. Schroeder, *Vorlesungen ueber die Algebra der Logik* (Leipzig 1890), Vol. 1, §1. Schroeder must however be criticized for not distinguishing two fundamentally different relations; the relation of an object to a concept it falls under, and the subordination of one concept to another. His remarks on the *Vollwurzel* are likewise open to objection. Schroeder's symbol ∈ does not simply take the place of the copula.

‡ Cf. my *Foundations* §66, footnote.

no other than Venus'; what was previously implicit in the single word 'is' is here set forth in four separate words, and in 'is no other than' the word 'is' now really is the mere copula. What is predicated here is thus not *Venus* but *no other than Venus*. These words stand for a concept; admittedly only one object falls under this, but such a concept must still always be distinguished from the object.* We have here a word 'Venus' that can never be a proper predicate although it can form part of a predicate. The reference† of this word is thus something that can never occur as a concept, but only as an object. Kerry, too, would probably not wish to dispute that there is something of this kind. But this would mean admitting a distinction, which it is very important to recognize, between what can occur only as an object, and everything else. And this distinction would not be effaced even if it were true, as Kerry thinks it is, that there are concepts that can also be objects.

There are, indeed, cases that seem to support his view. I myself have indicated (in *Foundations*, § 53, *ad fin.*) that a concept may fall under a higher concept – which, however, must not be confused with one concept's being subordinate to another. Kerry does not appeal to this; instead, he gives the following example: 'the concept "horse" is a concept easily attained,' and thinks that the concept 'horse' is an object, in fact one of the objects that fall under the concept 'concept easily attained.' quite so; the three words 'the concept "horse"' do designate an object but on that very account they do not designate a concept, as I am using the word. This is in full accord with the criterion I gave – that the singular definite article always indicates an object, whereas the indefinite article accompanies a concept-word.‡

Kerry holds that no logical rules can be based on linguistic distinctions; but my own way of doing this is something that nobody can avoid who lays down such rules at all; for we cannot come to an understanding with one another apart from language, and so in the end we must always rely on other people's understanding words, inflexions, and sentence-construction in essentially the same way as ourselves. As I said before, I was not trying to give a definition, but only hints; and to this end I appealed to the general feeling for the German language. It is here very much to my advantage that there is such good accord between the linguistic distinction and the real one. As regards the indefinite article there are probably no exceptions to our rule at all for us to remark apart from obsolete formulas like *Ein edler Rath* ['Councillor']. The matter is not so simple for the definite article, especially in the plural; but then my criterion does not relate to this case. In the singular, so far as I can see, the matter is doubtful only when a singular takes the place of a plural, as in the sentence 'the Turk besieged Vienna,' 'the horse is a four-legged animal.' These cases are so easily recognizable as special ones that the

---

* Cf. my *Foundations* §51.
† Cf. my paper, 'On Sense and Reference.'
‡ *Foundations*, §51; §66, footnote; §68, footnote on p.80.

value of our rule is hardly impaired by their occurrence. It is clear that in the first sentence 'the Turk' is the proper name for a people. The second sentence is probably best regarded as expressing a universal judgement, say 'all horses are four-legged animals' or 'all properly constituted horses are four-legged animals'; these will be discussed later.* Kerry calls my criterion unsuitable; for surely, he says, in the sentence 'the concept that I am now talking about is an individual concept' the name composed of the first eight words stands for a concept; but he is not taking the word 'concept' in my sense, and it is not in what I have laid down that the contradiction lies. But nobody can require that my mode of expression shall agree with Kerry's.

It must indeed be recognized that here we are confronted by an awkwardness of language, which I admit cannot be avoided, if we say that the concept *horse* is not a concept,† whereas, e.g., the city of Berlin is a city, and the volcano Vesuvius is a volcano. Language is here in a predicament that justifies the departure from custom. The peculiarity of our case is indicated by Kerry himself, by means of the quotation-marks around 'horse'; I use italics to the same end. There was no reason to mark out the words 'Berlin' and 'Vesuvius' in a similar way. In logical discussions one quote often needs to assert something about a concept, and to express this in the form usual for such assertions – viz. to make what is asserted of the concept into the content of the grammatical predicate. Consequently, one would expect that the reference of the grammatical subject would be the concept; but the concept as such cannot play this part, in view of its predicative nature; it must first be converted into an object,‡ or, speaking more precisely, represented by an object. We designate this object by prefixing the words 'the concept'; e.g.:

---

* Nowadays people seem inclined to exaggerate the scope of the statement that different linguistic expressions are never completely equivalent, that a word can never be exactly translated into another language. One might perhaps go even further, and say that the same word is never taken in quite the same way even by men who share a language. I will not enquire as to the measure of truth in these statements; I would only emphasize that nevertheless different expressions quite often have something in common, which I call the sense, or, in the special case of sentences, the thought. In other words, we must not fail to recognize that the same sense, the same thought, may be variously expressed; thus the difference does not here concern the sense, but only the apprehension, shading, or colouring of the thought, and is irrelevant for logic. It is possible for one sentence to give no more and no less information than another; and, for all the multiplicity of languages, mankind has a common stock of thoughts. If all transformation of the expression were forbidden on the plea that this would alter the content as well, logic would simply be crippled; for the task of logic can hardly be performed without trying to recognize the thought in its manifold guises. Moreover, all definitions would then have to be rejected as false.

† A similar thing happens when we say as regards the sentence 'this rose is red': The grammatical predicate 'is red' belongs to the subject 'this rose.' Here the words 'The grammatical predicate "is red"' are not a grammatical predicate but a subject. By the very act of explicitly calling it a predicate, we deprive it of this property.

‡ Cf. my *Foundations*, p. x.

'The concept *man* is not empty.'

Here the first three words are to be regarded as a proper name,* which can no more be used predicatively than 'Berlin' or 'Vesuvius.' When we say 'Jesus falls under the concept *man*,' then, setting aside the copula, the predicate is:

'someone falling under the concept *man*'

and this means the same as:

'a man.'

But the phrase

'the concept *man*'

is only part of this predicate.

Somebody might urge, as against the predicative nature of the concept, that nevertheless we speak of a subject-concept. But even in such cases, e.g. in the sentence

'all mammals have red blood'

we cannot fail to recognize the predicative nature† of the concept; for we could say instead:

'whatever is a mammal has red blood'

or:

'if anything is a mammal, then it has red blood.'

When I wrote my *Foundations of Arithmetic*, I had not yet made the distinction between sense and reference;‡ and so, under the expression 'a possible content of judgement,' I was combining what I now designate by the distinctive words 'thought' and 'truth-value.' Consequently, I no longer entirely approve of the explanation I then gave (op. cit., p. 77), as regards its wording; my view is, how-

---

* I call anything a proper name if it is a sign for an object.
† What I call here the predicative nature of the concept is just a special case of the need of supplementation, the 'unsaturatedness,' that I gave as the essential feature of a function in my work *Function and Concept* (Jena, 1891). It was there scarcely possible to avoid the expression 'the function $f(x)$,' although there too the difficulty arose that what this expression stands for is not a function.
‡ Cf. my essay 'On Sense and Reference.'

ever, still essentially the same. We may say in brief, taking 'subject' and 'predicate' in the linguistic sense: A concept is the reference of a predicate; an object is something that can never be the whole reference of a predicate, but can be the reference of a subject. It must here be remarked that the words 'all,' 'any,' 'no,' some,' are prefixed to concept-words. In universal and particular affirmative and negative sentences, we are expressing relations between concepts; we use these words to indicate the special kind of relation. They are thus, logically speaking, not to be more closely associated with the concept-words that follow them, but are to be related to the sentence as a whole. It is easy to see this in the case of negation. If in the sentence

'all mammals are land-dwellers'

the phrase 'all mammals' expressed the logical subject of the predicate *are land-dwellers*, then in order to negate the whole sentence we should have to negate the predicate: 'are not land-dwellers.' Instead we must put the 'not' in front of 'all'; from which it follows that 'all' logically belongs with the predicate. On the other hand, we do negate the sentence 'The concept *mammal* is subordinate to the concept *land-dweller*.'

If we keep it in mind that in my way of speaking expressions like 'the concept *F*' designate not concepts but objects, most of Kerry's objections already collapse. If he thinks that I have identified concept and extension of concept, he is mistaken; I merely expressed my view that in the expression 'the number that applies to the concept *F* is the extension of the concept *like-numbered to the concept F*' the words 'extension of the concept' could be replaced by 'concept.' Notice carefully that here the word 'concept' is combined with the definite article. Besides, this was only an incidental remark; I did not base anything upon it.

Thus Kerry does not succeed in filling the gap between concept and object. Someone might attempt, however, to make use of my own statements in this sense. I have said that to assign a number involves an assertion about a concept;* I speak of properties asserted of a concept, and I allow that a concept may fall under a higher one.† I have called existence a property of a concept. How I mean this to be taken is best made clear by an example. In the sentence 'there is at least one square root of 4,' we have an assertion, not about (say) the definitive number 2, nor about −2, but about a concept, *square root of 4*; viz. that it is not empty. But if I express the same thought thus: 'The concept *square root of 4* is realized,' then the first six words form the proper name of an object, and it is about this object that something is asserted. But notice carefully that what is asserted here is not the same thing as was asserted about the concept. This will be surprising only to somebody who fails

---

* *Foundations*, § 46.
† *Foundations*, § 53.

to see that a thought can be split up in many ways, so that now one thing, now another, appears as subject or predicate. The thought itself does not yet determine what is to be regarded as the subject. If we say 'the subject of this judgment,' we do not designate anything definite unless at the same time we indicate a definite kind of analysis; as a rule, we do this in connexion with a definite wording. But we must never forget that different sentences may express the same thought. For example, the thought we are considering could also be taken as an assertion about the number 4:

'The number 4 has the property that there is something of which it is the square.'

Language has means of presenting now one, now another, part of the thought as the subject; one of the most familiar is the distinction of active and passive forms. It is thus not impossible that one way of analysing a given thought should make it appear as a singular judgment; another, as a particular judgment; and a third, as a universal judgment. It need not then surprise us that the same sentence may be conceived as an assertion about a concept and also as an assertion about an object; only we must observe that what is asserted is different. In the sentence 'there is at least one square root of 4' it is impossible to replace the words 'square root of 4' by 'the concept *square root of 4*'; i.e. the assertion that suits the concept does not suit the object. Although our sentence does not present the concept as a subject, it asserts something about it; it can be regarded as expressing the fact that a concept falls under a higher one.* But this does not in any way efface the distinction between object and concept. We see to begin with that in the sentence 'there is at least one square root of 4' the predicative nature of the concept is not belied; we could say 'there is something that has the property of giving the result 4 when multiplied by itself.' Hence what is here asserted about a concept can never be asserted about an object; for a proper name can never be a predictive expression, though it can be part of one. I do not want to say it is false to assert about an object what is asserted here about a concept; I want to say it is impossible, senseless, to do so. The sentence 'there is Julius Caesar' is neither true nor false but senseless; the sentence 'there is a man whose name is Julius Caesar' has a sense, but here again we have a concept, as the indefinite article shows. We get the same thing in the sentence 'there is only one Vienna.' We must not let ourselves be deceived because language often uses the same word now as a proper name, now as a concept-word; in our example, the numeral indicates that we have the latter; 'Vienna' is here a concept-word, like 'metropolis.' Using it in this sense, we may say: 'Trieste is no Vienna.' If, on the other hand, we substitute 'Julius Caesar' for the proper name formed by the first six words of the sentence 'the concept *square*

---

* In my *Foundations* I called such a concept a second-order concept; in my work *Function and Concept* I called it a second-level concept, as I shall do here.

root of 4 is realized,' we get a sentence that has a sense but is false; for the assertion that something is realized (as the word is being taken here) can be truly made only about a quite special kind of objects, viz. such as can be designated by proper names of the form 'the concept *F*.' Thus the words 'the concept *square root of 4*' have an essentially different behaviour, as regards possible substitutions, from the words 'square root of 4' in our original sentence; i.e. the reference of the two phrases is essentially different.*

What has been shown here in one example holds good generally; the behaviour of the concept is essentially predicative, even where something is being asserted about it; consequently it can be replaced there only by another concept, never by an object. Thus the assertion that is made about a concept does not suit an object. Second-level concepts, which concepts fall under, are essentially different from first-level concepts, which objects fall under. The relation of an object to a first-level concept that it falls under is different from the (admittedly similar) relation of a first-level to a second-level concept. (To do justice at once to the distinction and to the similarity, we might perhaps say: An object falls *under* a first-level concept; a concept falls *within* a second-level concept.) The distinction of a concept and object thus still holds, with all its sharpness.

With this there hangs together what I have said (*Foundations*, §53) about my usage of the words 'property' and 'mark'; Kerry's discussion gives me occasion to revert once more to this. The words serve to signify relations, in sentences like 'Φ is a property of Γ' and 'Φ is a mark of Ω.' In my way of speaking, a thing can be at once a property and a mark, but not of the same thing. I call the concepts under which an object falls its properties; thus

'to be Φ is a property of Γ'

is just another way of saying:

'Γ falls under the concept of a Φ.'

If the object Γ has the properties Φ, X, and Ψ, I may combine them into Ω; so that it is the same thing if I say that Γ has the property Ω, or, that Γ has the properties Φ, X, and Ψ. I then call Φ, X, and Ψ marks of the concept Ω, and, at the same time, properties of Γ. It is clear that the relations of Φ to Γ and to Ω are quite different, and that consequently different terms are required. Γ falls under the concept Φ; but Ω, which is itself a concept, cannot fall under the first-level concept Φ; only to a second-level concept could it stand in a similar relation. Ω is, on the other hand, subordinate to Φ.

---

* Cf. my essay 'On Sense and Reference.'

Let us consider an example. Instead of saying:

> '2 is a positive number' and
> '2 is a whole number' and
> '2 is less than 10'

we may also say

> '2 is a positive whole number less than 10.'

Here

> *to be a positive number,*
> *to be a whole number,*
> *to be less than 10,*

appear as properties of the object 2, and also as marks of the concept

> *positive whole number less than 10.*

This is neither positive, nor a whole number, nor less than 10. It is indeed subordinate to the concept *whole number*, but does not fall under it.

Let us now compare with this what Kerry says in his second article. 'By the number 4 we understand the result of additively combining 3 and 1. The concept object here occurring is the numerical individual 4; a quite definite number in the natural number-series. This object obviously bears just the marks that are named in its concept, and no others besides – provided we refrain, as we surely must, from counting as *propria* of the object its infinitely numerous relations to all other individual numbers; "the" number 4 is likewise the result of additively combining 3 and 1.'

We see at once that my distinction between property and mark is here quite slurred over. Kerry distinguishes here between the number 4 and 'the' number 4. I must confess that this distinction is incomprehensible to me. The number 4 is to be a concept; 'the' number 4 is to be a concept-object, and none other than the numerical individual 4. It needs no proof that what we have here is not my distinction between concept and object. It almost looks as though what was floating (though very obscurely) before Kerry's mind were my distinction between the sense and the reference of the words 'the number 4.' But it is only of the reference of the words that we can say: this is the result of additively combining 3 and 1.

Again, how are we to take the word 'is' in the sentences 'the number 4 is the result of additively combining 3 and 1' and ' "the" number 4 is the result of additively combining 3 and 1'? Is it a mere copula, or does it help to express a logical equation? In the first case, 'the' would have to be left out before 'result,' and the sentences would go like this:

> 'The number 4 is a result of additively combining 3 and 1';
> '"The" number 4 is a result of additively combining 3 and 1.'

In that case, the objects that Kerry designates by

> 'the number 4' and ' "the" number 4'

would both fall under the concept

> *result of additively combining 3 and 1.*

And then the only question would be what difference there was between these objects. (I am here using the words 'object' and 'concept' in my accustomed way.) I should express as follows what Kerry is apparently trying to say:

> 'The number 4 has those properties, and those alone,
> which are marks of the concept: *result of additively combining 3 and 1.*'

I should then express as follows the sense of the first of our two sentences:

> 'To be a number 4 is the same as being a result
> of additive combination of 3 and 1.'

In that case, what I conjectured just now to have been Kerry's intention could also be put thus:

> 'The number 4 has those properties, and those alone,
> which are marks of the concept *a number 4.*'

(We need not here decide whether this is true.) The inverted commas around the definite article in the words ' "the" number 4' could in that case be omitted.

But in these attempted interpretations we have assumed that in at least one of the two sentences the definite articles in front of 'result' and 'number 4' were inserted only by an oversight. If we take the words as they stand, we can only regard them as having the sense of a logical equation, like:

> 'The number 4 is none other than the result
> of additively combining 3 and 1.'

The definite article in front of 'result' is here logically justified only if it is known (i) that there is such a result; (ii) that there is not more than one. In that case, the phrase designates an object, and is to be regarded as a proper name. If both our sentences were to be regarded as logical equations, then, since their right sides are identical, it would follow from them that the number 4 is 'the' number 4, or, if you prefer, that the number 4 is no other than 'the' number 4; and so Kerry's distinction would have been proved untenable. However, it is not my present task to point out contradictions in his exposition; his way of taking the words 'object' and 'concept' is not properly my concern here. I am only trying to set my own usage of these words in a clearer light, and incidentally show that in any case it differs from his, whether that is consistent or not.

I do not at all dispute Kerry's right to use the words 'concept' and 'object' in his own way, if only he would respect my equal right, and admit that with my use of terms I have got hold of a distinction of the highest importance. I admit that there is a quite peculiar obstacle in the way of an understanding with my reader. By a kind of necessity of language, my expressions, taken literally, sometimes miss my thought; I mention an object, when what I intend is a concept. I fully realize that in such cases I was relying upon a reader who would be ready to meet me half-way – who does not begrudge a pinch of salt.

Somebody may think that this is an artificially created difficulty; that there is no need at all to take account of such an unmanageable thing as what I call a concept; that one might, like Kerry, regard an object's falling under a concept as a relation, in which the same thing could occur now as object, now as concept. The words 'object' and 'concept' would then serve only to indicate the different positions in the relation. This may be done; but anybody who thinks the difficulty is avoided this way is very much mistaken; it is only shifted. For not all the parts of a thought can be complete; at least one must be 'unsaturated,' or predicative; otherwise they would not hold together. For example, the sense of the phrase 'the number 2' does not hold together with that of the expression 'the concept *prime number*' without a link. We apply such a link in the sentence 'the number 2 falls under the concept *prime number*'; it is contained in the words 'falls under,' which need to be completed in two ways – by a subject and an accusative; and only because their sense is thus 'unsaturated' are they capable of serving as a link. Only when they have been supplemented in this twofold respect do we get a complete sense, a thought. I say that such words or phrases stand for a relation. We now get the same difficulty for the relation that we were trying to avoid for the concept. For the words 'the relation of an object to the concept it falls under' designate not a relation but an object; and the three proper names 'the number 2,' 'the concept *prime number*,' 'the relation of an object to a concept it falls under,' hold aloof from one another just as much as the first two do by themselves; however we put them together, we get no sentence. It is thus easy for us to see that the difficulty arising from the 'unsaturatedness' of one part of the thought can indeed be shifted, but not avoided. 'Complete' and 'unsaturated' are of course only figures of speech; but all that I wish or am able to do here is to give hints.

It may make it easier to come to an understanding if the reader compares my work *Function and Concept*. For over the question what it is that is called a function in Analysis, we come up against the same obstacle; and on thorough investigation it will be found that the obstacle is essential, and founded on the nature of our language; that we cannot avoid a certain inappropriateness of linguistic expression; and that there is nothing for it but to realize this and always take it into account.

7
ON SENSE AND REFERENCE

EQUALITY* gives rise to challenging questions which are not altogether easy to answer. Is it a relation? A relation between objects, or between names or signs of objects? In my *Conceptual Notation* I assumed the latter. The reasons which seem to favour this are the following: $a = a$ and $a = b$ are obviously statements of differing cognitive value; $a = a$ holds *a priori* and, according to Kant, is to be labelled analytic, while statements of the form $a = b$ often contain very valuable extensions of our knowledge and cannot always be established *a priori*. The discovery that the rising sun is not new every morning, but always the same, was one of the most fertile astronomical discoveries. Even to-day the identification of a small planet or a comet is not always a matter of course. Now if we were to regard equality as a relation between that which the names 'a' and 'b' designate, it would seem that $a = b$ could not differ from $a = a$ (i.e. provided $a = b$ is true). A relation would thereby be expressed of a thing to itself, and indeed one in which each thing stands to itself but to no other thing. What is intended to be said by $a = b$ seems to be that the signs or names 'a' and 'b' designate the same thing, so that those signs themselves would be under discussion; a relation between them would be asserted. But this relation would hold between the names or signs only in so far as they named or designated something. It would be mediated by the connexion of each of the two signs with the same designated thing. But this is arbitrary. Nobody can be forbidden to use any arbitrarily producible event or object as a sign for something. In that case the sentence $a = b$ would no longer refer to the subject matter, but only to its mode of designation; we would express no proper knowledge by its means. But in many cases this is just what we want to do. If the sign 'a' is distinguished from the sign 'b' only as object (here, by means of its shape), not as sign (i.e. not by the manner in which it designates something), the cognitive value of $a = a$ becomes essentially equal to that of $a = b$, provided $a = b$ is true. A difference can arise only if the difference between the signs corresponds to a difference in the mode of presentation of

---

* I use this word strictly and understand '$a=b$' to have the sense of 'a is the same as b' or a and b coincide.'

that which is designated. Let *a*, *b*, *c* be the lines connecting the vertices of a triangle with the midpoints of the opposite sides. The point of intersection of *a* and *b* is then the same as the point of intersection of *b* and *c*. So we have different designations for the same point, and these names ('point of intersection of *a* and b ,' 'point of intersection of *b* and *c*') likewise indicate the mode of presentation; and hence the statement contains actual knowledge.

It is natural, now, to think of there being connected with a sign (name, combination of words, letter), besides that to which the sign refers, which may be called the reference of the sign, also what I should like to call the *sense* of the sign, wherein the mode of presentation is contained. In our example, accordingly, the reference of the expressions 'the point of intersection of *a* and *b*' and 'the point of intersection of *b* and *c*' would be the same, but not their senses. The reference of 'evening star' would be the same as that of 'morning star,' but not the sense.

It is clear from the context that by 'sign' and 'name' I have here understood any designation representing a proper name, which thus has as its reference a definite object (this word taken in the widest range), but not a concept or a relation, which shall be discussed further in another article.* The designation of a single object can also consist of several words or other signs. For brevity, let every such designation be called a proper name.

The sense of a proper name is grasped by everybody who is sufficiently familiar with the language or totality of designations to which it belongs;† but this serves to illuminate only a single aspect of the reference, supposing it to have one. Comprehensive knowledge of the reference would require us to be able to say immediately whether any given sense belongs to it. To such knowledge we never attain.

The regular connexion between a sign, its sense, and its reference is of such a kind that to the sign there corresponds a definite sense and to that in turn a definite reference, while to a given reference (an object) there does not belong only a single sign. The same sense has different expressions in different languages or even in the same language. To be sure, exceptions to this regular behaviour occur. To every expression belonging to a complete totality of signs, there should certainly correspond a definite sense; but natural languages often do not satisfy this condition, and one must be content if the same word has the same sense in the same context. It may perhaps be granted that every grammatically well-formed expression representing a proper name always has a sense. But this is not to say

---

* Editor's note: The reference is to [6].
† In the case of an actual proper name such as 'Aristotle' opinions as to the sense may differ. It might, for instance, be taken to be the following: the pupil of Plato and teach of Alexander the Great. Anybody who does this will attach another sense to the sentence 'Aristotle was born in Stagira' than will a man who takes as the sense of the name: the teacher of Alexander the Great who was born in Stagira. So long as the reference remains the same, such variations of sense may be tolerated, although they are to be avoided in the theoretical structure of a demonstrative science and ought not to occur in a perfect language.

that to the sense there also corresponds a reference. The words 'the celestial body most distant from the Earth' have a sense, but it is very doubtful if they also have a reference. The expression 'the least rapidly convergent series' has a sense; but it is known to have no reference, since for every given convergent series, another convergent, but less rapidly convergent, series can be found. In grasping a sense, one is not certainly assured of a reference.

If words are used in the ordinary way, what one intends to speak of is their reference. It can also happen, however, that one wishes to talk about the words themselves or their sense. This happens, for instance, when the words of another are quoted. One's own words then first designate words of the other speaker, and only the latter have their usual reference. We then have signs of signs. In writing, the words are in this case enclosed in quotation marks. Accordingly, a word standing between quotation marks must not be taken as having its ordinary reference.

In order to speak of the sense of an expression 'A' one may simply use the phrase 'the sense of the expression "A"'. In reported speech one talks about the sense, e.g., of another person's remarks. It is quite clear that in this way of speaking words do not have their customary reference but designate what is usually their sense. In order to have a short expression, we will say: In reported speech, words are used *indirectly* or have their *indirect* reference. We distinguish accordingly the *customary* from the *indirect* reference of a word; and its *customary* sense from its *indirect* sense. The indirect reference of a word is accordingly its customary sense. Such exceptions must always be borne in mind if the mode of connexion between sign, sense, and reference in particular cases is to be correctly understood.

The reference and sense of a sign are to be distinguished from the associated idea. If the reference of a sign is an object perceivable by the senses, my idea of it is an internal image,* arising from memories of sense impressions which I have had and acts, both internal and external, which I have performed. Such an idea is often saturated with feeling; the clarity of its separate parts varies and oscillates. The same sense is not always connected, even in the same man, with the same idea. The idea is subjective: one man's idea is not that of another. There result, as a matter of course, a variety of differences in the ideas associated with the same sense. A painter, a horseman, and a zoologist will probably connect different ideas with the name 'Bucephalus.' This constitutes an essential distinction between the idea and the sign's sense, which may be the common property of many and therefore is not a part of a mode of the individual mind. For one can hardly deny that mankind has a common store of thoughts which is transmitted from one generation to another.†

---

* We can include with ideas the direct experiences in which sense-impressions and acts themselves take the place of the traces which they have left in the mind. The distinction is unimportant for our purpose, especially since memories of sense-impressions and acts always help to complete the perceptual image. One can also understand direct experience as including any object, in so far as it is sensibly perceptible or spatial.

† Hence it is inadvisable to use the word 'idea' to designate something so basically different.

In the light of this, one need have no scruples in speaking simply of *the* sense, whereas in the case of an idea one must, strictly speaking, add to whom it belongs and at what time. It might perhaps be said: Just as one man connects this idea, and another that idea, with the same word, so also one man can associated this sense and another that sense. But there still remains a difference in the mode of connexion. They are not prevented from grasping the same sense; but they cannot have the same idea. *Si duo idem faciunt, non est idem.* If two persons picture the same thing, each still has his own idea. It is indeed sometimes possible to establish differences in the ideas, or even in the sensations, of different men; but an exact comparison is not possible, because we cannot have both ideas together in the same consciousness.

The reference of a proper name is the object itself which we designate by its means; the idea, which we have in that case, is wholly subjective; in between lies the sense, which is indeed no longer subjective like the idea, but is yet not the object itself. The following analogy will perhaps clarify these relationships. Somebody observes the Moon through a telescope. I compare the Moon itself to the reference; it is the object of the observation, mediated by the real image projected by the object glass in the interior of the telescope, and by the retinal image of the observer. The former I compare to the sense, the latter is like the idea or experience. The optical image in the telescope is indeed one-sided and dependent upon the standpoint of observation; but it is still objective, inasmuch as it can be used by several observers. At any rate it could be arranged for several to use it simultaneously. But each one would have his own retinal image. On account of the diverse shapes of the observers' eyes, even a geometrical congruence could hardly be achieved, and an actual coincidence would be out of the question. This analogy might be developed still further, by assuming A's retinal image made visible to B; or A might also see his own retinal image in a mirror. In this way we might perhaps show how an idea can itself be taken as an object, but as such is not for the observer what it directly is for the person having the idea. But to pursue this would take us too far afield.

We can now recognize three levels of difference between words, expressions, or whole sentences. The difference may concern at the most the ideas, or the sense but not the reference, or, finally, the reference as well. With respect to the first level, it is to be noted that, on account of the uncertain connexion of ideas with words, a difference may hold for one person, which another does not find. The difference between a translation and the original text should properly not overstep the first level. To the possible difference here belong also the colouring and shading which poetic eloquence seeks to give to the sense. Such colouring and shading are not objective, and must be evoked by each hearer or reader according to the hints of the poet or the speaker. Without some affinity in human ideas art would certainly be impossible; but it can never be exactly determined how far the intentions of the poet are realized.

In what follows there will be no further discussion of ideas and experiences;

they have been mentioned here only to ensure that the idea aroused in the hearer by a word shall not be confused with its sense or its reference.

To make a short and exact expression possible, let the following phraseology be established:

A proper name (word, sign, sign combination, expression) *expresses* its sense, *stands for* or *designates* its reference. By means of a sign we express its sense and designate its reference.

Idealists or sceptics will perhaps long since have objected: 'You talk, without further ado, of the Moon as an object; but how do you know that the name 'the Moon' has any reference?' I reply that when we say 'the Moon,' we do not intend to speak of our idea of the Moon, nor are we satisfied with the sense alone, but we presuppose a reference. To assume that in the sentence 'The Moon is smaller than the Earth' the idea of the Moon is in question, would be flatly to misunderstand the sense. If this is what the speaker wanted, he would use the phrase 'my idea of the Moon.' Now we can of course be mistaken in the presupposition, and such mistakes have indeed occurred. But the question whether the presupposition is perhaps always mistaken need not be answered here; in order to justify mention of the reference of a sign it is enough, at first, to point out our intention in speaking or thinking. (We must then add the reservation: provided such reference exists.)

So far we have considered the sense and reference only of such expressions, words, or signs as we have called proper names. We now inquire concerning the sense and reference for an entire declarative sentence. Such a sentence contains a thought.* Is this thought, now, to be regarded as its sense or its reference? Let us assume for the time being that the sentence has reference. If we now replace one word of the sentence by another having the same reference, but a different sense, this can have no bearing upon the reference of the sentence. Yet we can see that in such a case the thought changes; since, e.g., the thought in the sentence 'The morning star is a body illuminated by the Sun' differs from that in the sentence 'The evening star is a body illuminated by the Sun.' Anybody who did not know that the evening star is the morning star might hold the one thought to be true, the other false. The thought, accordingly, cannot be the reference of the sentence, but must rather be considered as the sense. What is the position now with regard to the reference? Have we a right even to inquire about it? Is it possible that a sentence as a whole has only a sense, but no reference? Have we a right even to inquire about it? Is it possible that a sentence as a whole has only a sense, but no reference? At any rate, one might expect that such sentences occur, just as there are parts of sentences having sense but no reference. And sentences which contain proper names without reference will be of this kind. The sentence 'Odysseus was set ashore at Ithaca while sound asleep' obviously has a sense. But since it is doubtful whether

---

* By a thought I understand not the subjective performance of thinking but its objective content, which is capable of being the common property of several thinkers.

the name 'Odysseus,' occurring therein, has reference, it is also doubtful whether the whole sentence has one. Yet it is certain, nevertheless, that anyone who seriously took the sentence to be true or false would ascribe to the name 'Odysseus' a reference, not merely a sense; for it is of the reference of the name that the predicate is affirmed or denied. Whoever does not admit the name has reference can neither apply nor withhold the predicate. But in that case it would be superfluous to advance to the reference of the name; one could be satisfied with the sense, if one wanted to go no further than the thought. If it were a question only of the sense of the sentence, the thought, it would be unnecessary to bother with the reference of a part of the sentence; only the sense, not the reference, of the part is relevant to the sense of the whole sentence. The thought remains the same whether 'Odysseus' has reference or not. The fact that we concern ourselves at all about the reference of a part of the sentence indicates that we generally recognize and expect a reference for the sentence itself. The thought loses value for us as soon as we recognize that the reference of one of its parts is missing. We are therefore justified in not being satisfied with the sense of a sentence, and in inquiring also as to its reference. But now why do we want every proper name to have not only a sense, but also a reference? Why is the thought not enough for us? Because, and to the extent that, we are concerned with its truth value. This is not always the case. In hearing an epic poem, for instance, apart from the euphony of the language we are interested only in the sense of the sentences and the images and feelings thereby aroused. The question of truth would cause us to abandon aesthetic delight for an attitude of scientific investigation. Hence it is a matter of no concern to us whether the name 'Odysseus,' for instance, has reference, so long as we accept the poem as a work of art.* It is the striving for truth that drives us always to advance from the sense to the reference.

We have seen that the reference of a sentence may always be sought, whenever the reference of its components is involved; and that in the case when and only when we are inquiring after the truth value.

We are therefore driven into accepting the *truth value* of a sentence as constituting its reference. By the truth value of a sentence I understand the circumstance that it is true or false. There are no further truth values. For brevity I call the one the True, the other the False. Every declarative sentence concerned with the reference of its words is therefore to be regarded as a proper name, and its reference, if it has one, is either the True or the False. These two objects are recognized, if only implicitly, by everybody who judges something to be true – and so even by a sceptic. The designation of the truth values as objects may appear to be an arbitrary fancy or perhaps mere play upon words, from which no profound consequences

---

* It would be desirable to have a special term for signs having only sense. If we name them, say, representations, the words of the actors on the stage would be representations; indeed the actor himself would be a representation.

could be drawn. What I mean by an object can be more exactly discussed only in connexion with concept and relation. I will reserve this for another article. But so much should already be clear, that in every judgement,* no matter how trivial, the step from the level of thoughts to the level of reference (the objective) has already been taken.

One might be tempted to regard the relation of the thought to the True not as that of sense to reference, but rather as that of subject to predicate. One can, indeed, say: "The thought, that 5 is a prime number, is true.' But closer examination shows that nothing more has been said than in the simple sentence '5 is a prime number.' The truth claim arises in each case from the form of the declarative sentence, and when the latter lacks its usual force, e.g., in the mouth of an actor upon the stage, even the sentence 'The thought that 5 is a prime number is true' contains only a thought, and indeed the same thought as the simple '5 is a prime number.' It follows that the relation of the thought to the True may not be compared with that of subject to predicate. Subject and predicate (understood in the logical sense) are indeed elements of thought; they stand on the same level for knowledge. By combining subject and predicate, one reaches only a thought, never passes from sense to reference, never from a thought to its truth value. One moves at the same level but never advances from one level to the next. A truth value cannot be a part of a thought, any more than, say, the Sun can, for it is not a sense but an object.

If our supposition that the reference of a sentence is its truth value is correct, the latter must remain unchanged when a part of the sentence is replaced by an expression having the same reference. And this is in fact the case. Leibniz gives the definition: '*Eadem sunt, quae sibi mutuo substitui possunt, salva veritate.*' What else but the truth value could be found, that belongs quite generally to every sentence if the reference of its components is relevant, and remains unchanged by substitutions of the kind in question?

If now the truth value of a sentence is its reference, then on the one hand all true sentences have the same reference and so, on the other hand, do all false sentences. From this we see that in the reference of the sentence all that is specific is obliterated. We can never be concerned only with the reference of a sentence; but again the mere thought alone yields no knowledge, but only the thought together with its reference, i.e. its truth value. Judgments can be regarded as advances from a thought to a truth value. Naturally this cannot be a definition. Judgment is something quite peculiar and incomparable. One might also say that judgments are distinctions of parts within truth values. Such distinction occurs by a return to the thought. To every sense belonging to a truth value there would correspond its own manner of analysis. However, I have here used the word 'part' in a special sense. I have in fact transferred the relation between the parts and the whole of the

---

* A judgment, for me is not the mere comprehension of a thought, but the admission of its truth.

sentence to its reference, by calling the reference of a word part of the reference of the sentence, if the word itself is part of the sentence. This way of speaking can certainly be attacked, because the whole reference and one part of it do not suffice to determine the remainder, and because the word 'part' is already used in another sense of bodies. A special term would need to be invented.

The supposition that the truth value of a sentence is its reference shall now be put to further test. We have found that the truth value of a sentence remains unchanged when an expression is replaced by another having the same reference: but we have not yet considered the case in which the expression to be replaced is itself a sentence. Now if our view is correct, the truth value of a sentence containing another part must remain unchanged when the part is replaced by another sentence having the same truth value. Exceptions are to be expected when the whole sentence or its part is direct or indirect quotation; for in such cases, as we have seen, the words do not have their customary reference. In direct quotation, a sentence designates another sentence, and in indirect quotation a thought.

We are thus led to consider subordinate sentences or clauses. These occur as parts of a sentence complex, which is, from the logical standpoint, likewise a sentence – a main sentence. But here we meet the question whether it is also true of the subordinate sentence that its reference is a truth value. Of indirect quotation we already know the opposite. Grammarians view subordinate clauses as representatives of parts of sentences and divide them accordingly into noun clauses, adjective clauses, adverbial clauses. This might generate the supposition that the reference of a subordinate clause was not a truth value but rather of the same kind as the reference of a noun or adjective or adverb – in short, of a part of a sentence, whose sense was not a thought but only a part of a thought. Only a more thorough investigation can clarify the issue. In so doing, we shall not follow the grammatical categories strictly, but rather group together what is logically of the same kind. Let us first search for cases in which the sense of the subordinate clause, as we have just supposed, is not an independent thought.

The case of an abstract noun clause, introduced by 'that,' includes the case of indirect quotation, in which we have seen the words to have their indirect reference coinciding with what is customarily their sense. In this case, then, the subordinate clause has for its reference a thought, not a truth value; as sense not a thought, but the sense of the words 'the thought, that ...,' which is only a part of the thought in the entire complex sentence. This happens after 'say,' 'hear,' 'be of the opinion,' 'be convinced,' 'conclude,' and similar words.* There is a different, and indeed somewhat complicated, situation after words like 'perceive,' 'know,' 'fancy,' which are to be considered later.

---

* In 'A lied in saying he had seen B,' the subordinate clause designates a thought which is said (1) to have been asserted by A (2) while A was convinced of its falsity.

That in the cases of the first kind the reference of the subordinate clause is in fact the thought can also be recognized by seeing that it is indifferent to the truth of the whole whether the subordinate clause is true or false. Let us compare, for instance, the two sentences 'Copernicus believed that the planetary orbits are circles' and 'Copernicus believed that the apparent motion of the sun is produced by the real motion of the Earth.' One subordinate clause can be substituted for the other without harm to the truth. The main clause and the subordinate clause together have as their sense only a single thought, and the truth of the whole includes neither the truth nor the untruth of the subordinate clause. In such cases it is not permissible to replace one expression in the subordinate clause by another having the same customary reference, but only by one having the same indirect reference, i.e. the same customary sense. If somebody were to conclude: The reference of a sentence is not its truth value, for in that case it could always be replaced by another sentence of the same truth value; he would prove too much; one might just as well claim that the reference of 'morning star' is not Venus, since one may not always say 'Venus' in place of 'morning star.' One has the right to conclude only that the reference of a sentence is not *always* its truth value, and that 'morning star' does not always stand for the planet Venus, viz. when the word has its indirect reference. An exception of such a kind occurs in the subordinate clause just considered which has a thought as its reference.

If one says 'It seems that ...' one means 'It seems to me that ...' or 'I think that ...' We therefore have the same case again. The situation is similar in the case of expressions such as 'to be pleased,' 'to regret,' 'to approve,' 'to blame,' 'to hope,' 'to fear.' If, toward the end of the battle of Waterloo, Wellington was glad that the Prussians were coming, the basis for his joy was a conviction. Had he been deceived, he would have been no less pleased so long as his illusion lasted; and before he became so convinced he could not have been pleased that the Prussians were coming – even though in fact they might have been already approaching.

Just as a conviction or a belief is the ground of a feeling, it can, as in inference, also be the ground of a conviction. In the sentence: 'Columbus inferred from the roundness of the Earth that he could reach India by travelling towards the west,' we have as the reference of the parts two thoughts, that the Earth is round, and that Columbus by travelling to the west could reach India. All that is relevant here is that Columbus was convinced of both, and that the one conviction was a ground for the other. Whether the Earth is really round, and whether Columbus could really reach India by travelling to the west, are immaterial to the truth of our sentence; but it is not immaterial whether we replace 'the Earth' by 'the planet which is accompanied by a moon whose diameter is greater than the fourth part of its own.' Here also we have the indirect reference of the words.

Adverbial final clauses beginning 'in order that' also belong here; for obviously the purpose is a thought; therefore: indirect reference for the words, subjunctive mood.

A subordinate clause with 'that' after 'command,' 'ask,' 'forbid,' would appear in direct speech as an imperative. Such a clause has no reference but only a sense. A command, a request, are indeed not thoughts, yet they stand on the same level as thoughts. Hence in subordinate clauses depending upon 'command,' 'ask,' etc., words have their indirect reference. The reference of such a clause is therefore not a truth value but a command, a request, and so forth.

The case is similar for the dependent question in phrases such as 'doubt whether,' 'not to know what.' It is easy to see that here also the words are to be taken to have their indirect reference. Dependent clauses expressing questions and beginning with 'who,' 'what,' 'where,' 'when,' 'how,' 'by what means,' etc., seem at times to approximate very closely to adverbial clauses in which words have their customary reference. These cases are distinguished linguistically [in German] by the mood of the verb. With the subjunctive, we have a dependent question and indirect reference of the words, so that a proper name cannot in general be replaced by another name of the same object.

In the cases so far considered the words of the subordinate clauses had their indirect reference, and this made it clear that the reference of the subordinate clause itself was indirect, i.e. not a truth value but a thought, a command, a request, a question. The subordinate clause could be regarded as a noun, indeed one could say: as a proper name of that thought, that command, etc., which it represented in the context of the sentence structure.

We now come to other subordinate clauses, in which the words do have their customary reference without however a thought occurring as sense and a truth value as reference. How this is possible is best made clear by examples.

<blockquote>Whoever discovered the elliptic form of the planetary orbits
died in misery.</blockquote>

If the sense of the subordinate clause were here a thought, it would have to be possible to express it also in a separate sentence. But this does not work, because the grammatical subject 'whoever' has no independent sense and only mediates the relation with the consequent clause 'died in misery.' For this reason the sense of the subordinate clause is not a complete thought, and its reference is Kepler, not a truth value. One might object that the sense of the whole does contain a thought as part, viz. that there was somebody who first discovered the elliptic form of the planetary orbits; for whoever takes the whole to be true cannot deny this part. This is undoubtedly so; but only because otherwise the dependent clause 'whoever discovered the elliptic form of the planetary orbits' would have no reference. If anything is asserted there is always an obvious presupposition that the simple or compound proper names used have reference. If one therefore asserts 'Kepler died in misery,' there is a presupposition that the name 'Kepler' designates something; but it does not follow that the sense of the sentence 'Kepler died in misery' contains the thought that the name 'Kepler' designates something. If this were the case the negation would have to run not

> Kepler did not die in misery

but

> Kepler did not die in misery,
> or the name 'Kepler' has no reference.

That the name 'Kepler' designates something is just as much a presupposition for the assertion

> Kepler died in misery

as for the contrary assertion. Now languages have the fault of containing expressions which fail to designate an object (although their grammatical form seems to qualify them for that purpose) because the truth of some sentences is a prerequisite. Thus it depends on the truth of the sentence:

There was someone who discovered the elliptic form of the planetary orbits

whether the subordinate clause

> Whoever discovered the elliptic form of the planetary orbits

really designates an object or only seems to do so while having in fact no reference. And thus it may appear as if our subordinate clause contained as a part of its sense the thought that there was somebody who discovered the elliptic form of the planetary orbits. If this were right the negation would run:

> Either whoever discovered the elliptic form of the planetary orbits
> did not die in misery or there was nobody who discovered
> the elliptic form of the planetary orbits.

This arises from an imperfection of language, from which even the symbolic language of mathematical analysis is not altogether free; even there combinations of symbols can occur that seem to stand for something but have (at least so far) no reference, e.g. divergent infinite series.. This can be avoided, e.g., by means of the special stipulation that divergent infinite series shall stand for the number 0. A logically perfect language (*Conceptual Notation*) should satisfy the conditions, that every expression grammatically well constructed as a proper name out of signs already introduced shall in fact designate an object, and that no new sign shall be introduced as a proper name without being secured a reference. The logic books contain warnings against logical mistakes arising from the ambiguity of expressions. I regard as no less pertinent a warning against apparent proper names having no reference. The history of mathematics supplies errors which have arisen in this way. This lends itself to demagogic abuse as easily as ambiguity – perhaps more easily. 'The will of the people' can serve as an example; for it is easy to establish that there is at ay rate no generally accepted reference for this expression. It is therefore by no means unimportant to eliminate the source of these mistakes, at least in science, once and for all. Then such objections as the one discussed above would become impossible, because it could never depend upon the truth of a thought whether a proper name had a reference.

With the consideration of these noun clauses may be coupled that of types of adjective and adverbial clauses which are logically in close relation to them.

Adjective clauses also serve to construct compound proper names, though, unlike noun clauses, they are not sufficient by themselves for this purpose. These adjective clauses are to be regarded as equivalent to adjectives. Instead of 'the square root of 4 which is smaller than 0,' one can also say 'the negative square root of 4.' We have here the case of a compound proper name constructed from the expression for a concept with the help of the singular definite article. This is at any rate permissible if the concept applies to one and only one single object.*

Expressions for concepts can be so constructed that marks of a concept are given by adjective clauses as, in our example, by the clause 'which is smaller than 0.' It is evident that such an adjective clause cannot have a thought as sense or a truth value as reference, any more than the noun clause could. Its sense, which can also be expressed in many cases by a single adjective, is only a part of a thought. Here, as in the case of the noun clause, there is no independent subject and therefore no possibility of reproducing the sense of the subordinate clause in an independent sentence.

Places, instants, stretches of time, are, logically considered, objects; hence the linguistic designation of a definite place, a definite instant, or a stretch of time is to be regarded as a proper name. Now adverbial clauses of place and time can be used for the construction of such a proper name in a manner similar to that which we have seen in the case of noun and adjective clauses. In the same way, expressions for concepts bringing in places, etc., can be constructed. It is to be noted here also that the sense of these subordinate clauses cannot be reproduced in an independent sentence, since an essential component, viz. the determination of place or time, is missing and is only indicated by a relative pronoun or a conjunction.†

---

* In accordance with what was said above, an expression of the kind in question must actually always be assured of reference, by means of a special stipulation, e.g. by the convention that 0 shall count as its reference, when the concept applies to no object or to more than one.
† In the case of these sentences, various interpretations are easily possible. The sense of the sentence, 'After Schleswig-Holstein was separated from Denmark, Prussia and Austria quarrelled' can also be rendered in the form 'After the separation of Schleswig-Holstein from Denmark, Prussia and Austria quarrelled.' In this version, it is surely sufficiently clear that the sense is not to be taken as having a part the thought that Schleswig-Holstein was once separated from Denmark, but that this is the necessary presupposition in order for the expression 'after the separation of Schleswig-Holstein from Denmark' to have any reference at all. To be sure, our sentence can also

In conditional clauses, also, there may usually be recognized to occur an indefinite indicator, having a similar correlate in the dependent clause. (We have already seen this occur in noun, adjective, and adverbial clauses.) In so far as each indicator refers to the other, both clauses together form a connected whole, which as a rule expresses only a single thought. In the sentence

> If a number is less than 1 and greater than 0,
> its square is less than 1 and greater than 0

the component in question is 'a number' in the conditional clause and 'its' in the dependent clause. It is by means of this very indefiniteness that the sense acquires the generality expected of a law. It is this which is responsible for the fact that the antecedent clause alone has no complete thought as its sense and in combination with the consequent clause expresses one and only one thought, whose parts are no longer thoughts. It is, in general, incorrect to say that in the hypothetical judgment two judgments are put in reciprocal relationship. If this or something similar is said, the word 'judgment' is used in the same sense and I have connected with the word 'thought,' so that I would use the formulation: 'A hypothetical thought establishes a reciprocal relationship between two thoughts.' This could be true only if an indefinite indicator is absent;* but in such a case there would also be no generality.

If an instant of time is to be indefinitely indicated in both conditional and dependent clauses, this is often achieved merely by using the present tense of the verb, which in such a case however does not indicate the temporal present. This grammatical form is then the indefinite indicator in the main and subordinate clauses. An example of this is: 'When the Sun is in the tropic of Cancer, the longest day in the northern hemisphere occurs.' Here, also, it is impossible to express the sense of the subordinate clause in a full sentence, because this sense is not a complete thought. If we say: 'The Sun is in the tropic of Cancer,' this would refer to our present time and thereby change the sense. Just as little is the sense of the main clause a thought; only the whole, composed of main and subordinate clauses, has such a sense. It may be added that several common components in the antecedent and consequent clauses may be indefinitely indicated.

---

be interpreted as saying that Schleswig-Holstein was once separated from Denmark. We then have a case which is to be considered later. In order to understand the difference more clearly, let us project ourselves into the mind of a Chinese who, having little knowledge of European history, believes it to be false that Schleswig-Holstein was ever separated from Denmark. He will take our sentence, in the first version, to be neither true nor false but will deny it to have any reference, on the ground of absence of reference for its subordinate clause. This clause would only apparently determine a time. If he interpreted our sentence in the second way, however, he would find a thought expressed in it which he would take to be false, beside a part which would be without reference for him.

* At times an explicit linguistic indication is missing and must be read off from the entire context.

It is clear that noun clauses with 'who' or 'what' and adverbial clauses with 'where', 'when,' 'wherever,' 'whenever' are often to be interpreted as having the sense of conditional clauses, e.g. 'who touches pitch, defiles himself.'

Adjective clauses can also take the place of conditional clauses. Thus the sense of the sentence previously used can be given in the form 'The square of a number which is less than 1 and greater than 0 is less than 1 and greater than 0.'

The situation is quite different if the common component of the two clauses is designated by a proper name. In the sentence:

> Napoleon, who recognized the danger to his right flank, himself led his guards against the enemy position

two thoughts are expressed:

1. Napoleon recognized the danger to his right flank
2. Napoleon himself led his guards against the enemy position.

When and where this happened is to be fixed only by the context, but is nevertheless to be taken as definitely determined thereby. If the entire sentence is uttered as an assertion, we thereby simultaneously assert both component sentences. If one of the parts is false, the whole is false. Here we have the case that the subordinate clause by itself has a complete thought as sense (if we complete it by indication of place and time). The reference of the subordinate clause is accordingly a truth value. We can therefore expect that it may be replaced, without harm to the truth value of the whole, by a sentence having the same truth value. This is indeed the case; but it is to be noticed that for purely grammatical reasons, its subject must be 'Napoleon,' for only then can it be brought into the form of an adjective clause belonging to 'Napoleon.' But if the demand that it be expressed in this form be waived, and the connexion be shown by 'and,' this restriction disappears.

Subsidiary clauses beginning with 'although' also express complete thoughts. This conjunction actually has no sense and does not change the sense of the clause but only illuminates it in a peculiar fashion.* We could indeed replace the conditional clause without harm to the truth of the whole by another of the same truth value; but the light in which the clause is placed by the conjunction might then easily appear unsuitable, as if a song with a sad subject were to be sung in a lively fashion.

In the last cases the truth of the whole included the truth of the component clauses. The case is different if a conditional clause expresses a complete thought by containing, in place of an indefinite indicator, a proper name or something which is to be regarded as equivalent. In the sentence

---

* Similarly in the case of 'but,' 'yet.'

> If the Sun has already risen, the sky is very cloudy

the time is the present, that is to say, definite. And the place is also to be thought of as definite. Here it can be said that a relation between the truth values of conditional and dependent clauses has been asserted, vis. such that the case does not occur in which the antecedent stands for the True and the consequent for the False. Accordingly, our sentence is true if the Sun has not yet risen, whether the sky is very cloudy or not, and also if the Sun has risen and the sky is very cloudy. Since only truth values are here in question, each component clause can be replaced by another of the same truth value without changing the truth value of the whole. To be sure, the light in which the subject then appears would usually be unsuitable; the thought might easily seem distorted; but this has nothing to do with its truth value. One must always take care not to clash with the subsidiary thoughts, which are however not explicitly expressed and therefore should not be reckoned in the sense. Hence, also, no account need to taken of their truth values.*

The simple cases have now been discussed. Let us review what we have learned.

The subordinate clause usually has for its sense not a thought, but only a part of one, and consequently no truth value as reference. The reason for this is either that the words in the subordinate clause have indirect reference, so that the reference, not the sense, of the subordinate clause is a thought; or else that, on account of the presence of an indefinite indicator, the subordinate clause is incomplete and expresses a thought only when combined with the main clause. It may happen, however, that the sense of the subsidiary clause is a complete thought, in which case it can be replaced by another of the same truth value without harm to the truth of the whole – provided there are no grammatical obstacles.

An examination of all the subordinate clauses which one may encounter will soon provide some which do not fit well into these categories. The reason, so far as I can see, is that these subordinate clauses have no such simple sense. Almost always, it seems, we connect with the main thoughts expressed by us subsidiary thoughts which, although not expressed, are associated with our words, in accordance with psychological laws, by the hearer. And since the subsidiary thought appears to be connected with our words of its own accord, almost like the main thought itself, we want it also to be expressed. The sense of the sentence is thereby enriched, and it may well happen that we have more simple thoughts than clauses. In many cases the sentence must be understood in this way, in others it may be doubtful whether the subsidiary thought belongs to the sense of the sentence or only accompanies it.† One might perhaps find that the sentence

---

* The thought of our sentence might also be expressed thus: 'Either the Sun has not risen yet or the sky is very cloudy' – which shows how this kind of sentence connexion is to be understood.

† This may be important for the question whether an assertion is a lie, or an oath a perjury.

> Napoleon, who recognized the danger to his right flank,
> himself led his guards against the enemy position

expresses not only the two thoughts shown above, but also the thought that the knowledge of the danger was the reason why he led the guards against the enemy position. One may in fact doubt whether this thought is merely slightly suggested or really expressed. Let the question be considered whether our sentence be false if Napoleon's decisions had already been made before he recognized the danger. If our sentence could be true in spite of this, the subsidiary thought should not be understood as part of the sense. One would probably decide in favour of this. The alternative would make for quite a complicated situation: We would have more simple thoughts than clauses. If the sentence

> Napoleon recognized the danger to his right flank

were now to be replaced by another having the same truth value, e.g.

> Napoleon was already more than 45 years old

not only would our first thought be changed, but also our third one. Hence the truth value of the latter might change – viz. if his age was not the reason for the decision to lead the guards against the enemy. This shows why clauses of equal truth value cannot always be substituted for one another in such cases. The clause expresses more through its connexion with another than it does in isolation.

Let us now consider cases where this regularly happens. In the sentence:

> Bebel mistakenly supposes that the return of Alsace-Lorraine
> would appease France's desire for revenge

two thoughts are expressed, which are not however shown by means of antecedent and consequent clauses, viz.:

(1) Bebel believes that the return of Alsace-Lorraine would appease France's desire for revenge
(2) the return of Alsace-Lorraine would not appease France's desire for revenge.

In the expression of the first thought, the words of the subordinate clause have their indirect reference, while the same words have their customary reference in the expression of the second thought. This shows that the subordinate clause in our original complex sentence is to be taken twice over, with different reference, standing once for a thought, once for a truth value. Since the truth value is not the whole reference of the subordinate clause, we cannot simply replace the latter by another of equal truth value. Similar considerations apply to expressions such as 'know,' 'discover,' 'it is known that.'

By means of a subordinate causal clause and the associated main clause we express several thoughts, which however do not correspond separately to the original clauses. In the sentence: 'Because ice is less dense than water, it floats on water' we have

(1) Ice is less dense than water;
(2) If anything is less dense than water, it floats on water;
(3) Ice floats on water.

The third thought, however, need not be explicitly introduced, since it is contained in the remaining two. On the other hand, neither the first and third nor the second and third combined would furnish the sense of our sentence. It can now be seen that our subordinate clause

because ice is less dense than water

expresses our first thought, as well as part of our second. This is how it comes to pass that our subsidiary clause cannot be simply replaced by another of equal truth value; for this would alter our second thought and thereby might well alter its truth value.

The situation is similar in the sentence

If iron were less dense than water, it would float on water.

Here we have the two thoughts that iron is not less dense than water, and that something floats on water if it is less dense than water. The subsidiary clause again expresses one thought and a part of the other.

If we interpret the sentence already considered

After Schleswig-Holstein was separated from Denmark,
Prussia and Austria quarrelled

in such a way that it expresses the thought that Schleswig-Holstein was once separated from Denmark, we have first this thought, and secondly the thought that at a time, more closely determined by the subordinate clause, Prussia and Austria quarrelled. Here also the subordinate clause expresses not only one thought but also a part of another. Therefore it may not in general be replaced by another of the same truth value.

It is hard to exhaust all the possibilities given by language; but I hope to have brought to light at least the essential reasons why a subordinate clause may not always be replaced by another of equal truth value without harm to the truth of the whole sentence structure. These reasons arise:

(1) when the subordinate clause does not stand for a truth value, inasmuch as it expresses only a part of a thought;
(2) when the subordinate clause does stand for a truth value but is not restricted to so doing, inasmuch as its sense includes one thought and part of another.

The first case arises:

(a) in indirect reference of words
(b) if a part of the sentence is only an indefinite indicator instead of a proper name.

In the second case, the subsidiary clause may have to be taken twice over, viz. once in its customary reference, and the other time in indirect reference; or the sense of a part of the subordinate clause may likewise be a component of another thought, which, taken together with the thought directly expressed by the subordinate clause, makes up the sense of the whole sentence.

It follows with sufficient probability from the foregoing that the cases where the subordinate clause is not replaceable by another of the same value cannot be brought in disproof of our view that a truth value is the reference of a sentence having a thought as its sense.

Let us return to our starting point.

When we found '$a = a$' and '$b = b$' to have different cognitive values, the explanation is that for the purpose of knowledge, the sense of the sentence, viz., the thought expressed by it, is no less relevant than its reference, i.e., its truth value. If now $a = b$, then indeed the reference of 'b' is the same as that of 'a,' and hence the truth value of '$a = b$' is the same as that of '$a = a$'. In spite of this, the sense of 'b' may differ from that of 'a,' and thereby the sense expressed in '$a = b$' differs from that of '$a = a$.' In that case the two sentences do not have the same cognitive value. If we understand by 'judgment' the advance from the thought to its truth value, as in the above paper, we can also say that the judgments are different.

8
WHAT IS A FUNCTION?

It is even now not beyond all doubt what the word 'function'* stands for in Analysis, although it has been in continual use for a long time. In definitions, we find two expressions constantly recurring, sometimes in combination and sometimes separately: 'mathematical expression' and 'variable.' We also notice a fluctuation usage: the name 'function' is given sometimes to what determines the mode of dependence, or perhaps to the mode of dependence itself, and sometimes to the dependent variable.

In recent times the word 'variable' is predominant in the definitions. But this is itself very much in need of explanation. Any variation occurs in time. Consequently Analysis would have to deal with a process in time, since it takes variables into consideration. But in fact it has nothing to do with time; its applicability to occurrences in time is irrelevant. There are also applications of Analysis to geometry; and here time is left quite out of account. This is one of the main difficulties, one that we encounter again and again when once we try to get away from examples to the root of the matter. For as soon as we try to mention a variable, we shall hit upon something that varies in time and thus does not belong to pure Analysis. And yet it must be possible to point to a variable that does not involve something alien to arithmetic, if variables are objects of Analysis at all.

If variation thus already raises a difficulty, we encounter a fresh one when we ask what varies. The answer one immediately gets is: a magnitude. Let us look for an example. We may call a rod a magnitude in respect of its being long. Any variation in the rod as regards its length, such as may result, e.g., from heating it, occurs in time; and neither rods nor lengths are objects of pure Analysis. This attempt to point to a variable magnitude within Analysis is a failure; and in just the same way, many others must fail; for the magnitudes of lengths, surfaces, angles, masses, are none of them objects of arithmetic. Among all magnitudes, only numbers belong to arithmetic; and it is just because this science leaves wholly indefinite what magnitudes were measured in particular cases so as to get numbers, that it admits of the most various

---

* Our discussion will be confined to functions of a single argument.

applications. Our question is, then: Are the variables of analysis variable numbers? What else could they be, if they are to belong to Analysis at all? But why is it that people hardly ever say 'variable number' but on the other hand often say 'variable magnitude'? The latter expression sounds more acceptable than 'variable number'; for as regards that there arises the doubt: are there variable numbers? Surely every number retains its properties, without varying. 'Of course,' someone may say, '3 and π are obviously invariable numbers, constants; but there are also variable numbers. For example, when I say "the number that gives the length of this rod in millimetres" I am naming a number; and this is variable, because the rod does not always keep the same length; so by using this expression I have designated a variable number.' Let us compare this example with the following one. 'When I say "the King of this realm" I am designating a man. Ten years ago the King of this realm was an old man; at present the King of this realm is a young man. So by using this expression I have designated a man who was an old man and is now a young man.' There must be something wrong here. The expression 'the King of this realm' does not designate any man at all, if the time is not mentioned; as soon, however, as mention of a time is added, it can designate one man unambiguously; but then this mention of a time is a necessary constituent of the expression, and we get a different expression if we mention a different time. Thus in our two sentences we just have not the same subject of predication. Similarly, the expression 'the number that gives the length of this rod in millimetres' does not designate any number at all if the time is not mentioned. If mention of a time is added, a number may thus be designated, e.g. 1,000; but then this is invariable. If a different time is mentioned, we get a different expression, which may thus also designate a different number, say, 1,001. If we say 'Half an hour ago the number that gave the length of this rod in millimetres was a cube; at present the number that gives the length of this rod in millimetres is not a cube,' we just have not got the same subject of predication. The number 1,000 has not somehow swollen up to 1,001, but has been replaced by it. Or is the number 1,000 perhaps the same as the number 1,001, only with a different expression on its face? If anything varies, we have in succession different properties, states, in the same object. If it were not the same one, we should have no subject of which we could predicate variation. A rod grows longer through being heated; while this is going on, it remains the same one. If instead it were taken away and replaced by a longer one, we could not say it had grown longer. A man grows older; but if we could not nevertheless recognize him as the same man, we should have nothing of which we could predicate growing older. Let us apply to this number. What remains the same when a number varies? Nothing! Hence a number does *not* vary; for we have nothing of which we could predicate the variation. A cube never turns into a prime number; an irrational number never becomes rational.

Thus there are no variable numbers; and this is confirmed by the fact that we have no proper names for variable numbers. We failed in our attempt to use the expression 'the number that gives the length of this rod in millimetres' as a designation of a variable number. But do we not use 'x,' 'y,' 'z' to designate variable numbers? This way of speaking is certainly employed; but these letters are not proper names of variable numbers in the way that '2' and '3' are proper names of constant numbers; for the numbers '2' and '3' differ in a specified way, but what is the difference between the variables that are said to be designated by 'x' and 'y'? We cannot say. We cannot specify what properties x has and what differing properties y has. If we associate anything with these letters at all, it is the same vague image for both of them. When apparent differences do show themselves, it is a matter of applications; but we are not here talking about these. Since we cannot conceive of each variable in its individual being, we cannot attach any proper names to variables.

Herr E. Czuber has attempted to avoid some of the difficulties I have mentioned.* In order to eliminate time, he defines the variable as an indefinite number. But are there indefinite numbers? Must numbers be divided into definite and indefinite? Are there indefinite men? Must not every object be definite? 'But is not the number n indefinite?' I am not acquainted with the number n. 'n' is not the proper name of any number, definite or indefinite. Nevertheless, we do sometimes say 'the number n.' How is this possible? Such an expression must be considered in a context. Let us take an example. 'If the number n is even, then $\cos n\pi = 1$.' Here only the whole has a sense, not the antecedent by itself nor the consequent by itself. The question whether the number n is even cannot be answered; no more can the question whether $\cos n\pi = 1$. For an answer to be given, 'n' would have to be the proper name of a number, and in that case this would necessarily be a definite one. We write the letter 'n' in order to achieve generality. This presupposes that, if we replace it by the name of a number, both antecedent and consequent receive a sense.

Of course we may speak of indefiniteness here; but here the word 'indefinite' is not an adjective of 'number,' but ['indefinitely'] is an adverb, e.g., of the verb 'to indicate.' We cannot say that 'n' designates an indefinite number, but we *can* say that it indicates numbers indefinitely. And so it is always when letters are used in arithmetic, except for the few cases (π, e, i) where they occur as proper names; but then they designate definite, invariable numbers. There are thus no indefinite numbers, and this attempt of Herr Czuber's is a failure.

The second deficiency that he tries to remedy is that we cannot conceive of any variable so as to distinguish it from others. He calls the totality of the values that a variable may assume, the range of the variable, and says: 'The variable x counts as having been defined when it can be determined as regards any assigned real number whether it belongs to the range or not.' It counts has having been defined; but *has* it? Since there are no indefinite numbers, it is impossible to define any indefinite number. The range is represented as distinctive for the

---

* *Vorlesungen uber Differential-une Integralrechnung* (Teubner, Leipzig) 1, §2.

variable; so with the same range we should have the same variable. Consequently in the equation '$y = x^2$' y would be the same variable as x if the range of x is that of positive numbers.

We must regard this attempt as having come to grief; in particular, the expression 'a variable assumes a value' is completely obscure. A variable is to be an indefinite number. Now how does an indefinite number set about assuming a number? for the value is obviously a number. Does, e.g., an indefinite man likewise assume a definite man? In other connexions, indeed, we say that an object assumes a property; here the number must play both parts; as an object it is called a variable or a variable magnitude, and as a property it is called a value. That is why people prefer the word 'magnitude' to the word 'number'; they have to deceive themselves about the fact that the variable magnitude and the value it is said to assume are essentially the same thing, that in this case we have *not* got an object assuming different properties in succession, and that therefore there can be no question of a variation.

As regards variables our results are as follows. Variable magnitudes may certainly be admitted, but do not belong to pure Analysis. Variable numbers do not exist. The word 'variable' thus has no justification in pure Analysis.

Now how do we get from the variable to the function? This will probably be done always in essentially the same way; so we follow Herr Czuber's way of putting it. He writes (§3): 'If every value of the real variable x that belongs to its range has correlated with it a definite number y, then in general y also is defined as a variable, and is called *a function of the real variable x*. This relation is expressed by an equation of the form $y = f(x)$.'

It is at once noticeable that y is called a definite number, whereas on the other hand, being a variable, it would have to be an indefinite number. y is neither a definite nor an indefinite number; but the sign 'y' is attached incorrectly to a plurality of numbers, and then afterwards he talks as if there were only a single number. It would be simpler and clearer to state the matter as follows. 'With every number of an x-range there is correlated a number. I call the totality of these numbers the y-range.' Here we certainly have a y-range, but we have no y of which we could say that it was a function of the real variable x.

Now the delimitation of the range appears irrelevant to the question what the function essentially is. Why could we not at once take the range to be the totality of real numbers, or the totality of complex numbers, including real numbers? The heart of the mater really lies in quite a different place, viz. hidden in the word 'correlated.' Now how do I tell whether the number 5 is correlated with the number 4? The question is unanswerable unless it is somehow completed. And yet with Herr Czuber's explanation it looks as though it were already determined, for any two numbers, whether the first is correlated with the second or not. Fortunately Herr Czuber adds the remark: 'The above definition involves no assertion as to the *law* of correlation, which is indicated in the most general way by the *characteristic f*; this can be set up in the most various ways.'

Correlation, then, takes place according to a law, and different laws of this sort can be thought of. In that case, the expression 'y is a function of x' has no sense, unless it is completed by mentioning the law of correlation. This is a mistake in the definition. And surely the law, which this definition treats as not being given, is really the main thing. We notice that now variability has dropped entirely out of sight; instead, generality comes into view, for that is what the word 'law' indicates.

Distinctions between laws of correlation will go along with distinctions between functions; and these cannot any longer be regarded as quantitative. If we just think of algebraic functions, the logarithmic function, elliptic functions, we convince ourselves immediately that here is a matter of qualitative differences; a further reason for not defining functions as variables. If they were variables, elliptic functions would be elliptic variables.

Our general way of expressing such a law of correlation is an equation, in which the letter 'y' stands on the right side whereas on the left there appears a mathematical expression consisting of numerals, mathematical signs, and the letter 'x,' e.g.:

$$'y = x^2 + 3x'$$

The function has indeed been defined as being such a mathematical expression. In recent times this concept has been found too narrow. However, this difficulty could easily be avoided by introducing new signs into the symbolic language of arithmetic. Another objection has more weight: viz. that a mathematical expression, as a group of signs, does not belong in arithmetic at all. The formalist theory, which regards signs as the subject-matter of this science, is one that I may well consider to be definitively refuted by my criticism in the second volume of my *Basic Laws of Arithmetic*. The distinction between sign and thing signified has not always been sharply made, so 'mathematical expression' (*expressio analytica*) has been half taken to mean also what the expression stands for. Now what does '$x^2 + 3x$' designate? Properly speaking, nothing at all; for the letter 'x' only indicates numbers, and does not designate them. If we replace 'x' by a numeral, we get an expression that designates a number, and so nothing new. Like 'x' itself, '$x^2 + 3x$' only indicates. This may be done for the sake of expressing generality, as in the sentences

$$'x^2 + 3x = x(x + 3)'$$
if $x > 0$ then '$x^2 + 3x > 0$.'

But now what has become of the function? It looks as though we could not take it to be either the mathematical expression itself or what the expression stands for. And yet we have not gone very far off the right track. Each of the expressions in 'sin 0,' 'sin 1,' 'sin 2' stands for a particular number; but we have a common constituent '*sin*,' and here we find a designation for the essential peculiarity of the sine-function. This '*sin*' perhaps corresponds to the '*f*' that Herr Czuber says indicates a law; and the transition from '*f*' to '*sin*,' just like that from '*a*' to '2,' is a transition from a sign that indicates to one that designates. In that case 'sin' would stand for a law. Of course that is not quite right. The law seems rather to be expressed in the equation 'y = sin *x*'; the symbol '*sin*' is only part of this, but the part that is distinctive for the essential peculiarity of the law. And surely we have here what we were looking for – the function? '*f*' too will then, strictly speaking, indicate a function. And here we come upon what distinguishes functions from numbers. '*Sin*' requires completion with a numeral, which, however, does not form part of the designation of the function. This holds good in general; the sign for a function is 'unsaturated'; it needs to be completed with a numeral, which we then call the argument-sign. We see this also with the root-sign, with the logarithm-sign. The functional sign cannot occur on one side of an equation by itself, but only when completed by a sign that designates or indicates a number. Now what does such a complex stand for, consisting of a functional sign and a numeral, e.g. 'sin 1,' '$\sqrt{1}$,' 'log 1'? A number each time. We thus get numerical signs composed of two dissimilar parts, an 'unsaturated' part being completed by the other one.

This need of completion may be made apparent by empty brackets, e.g. 'sin ( )' or '( )^2 + 3. ( ).' This is perhaps the most appropriate notation, and the one best calculated to avoid the confusion that arises from regarding the argument-sign as part of the functional sign; but it will probably not meet with any acceptance.* A letter may also be employed for this purpose. If we choose ξ, then 'sin ξ' and 'ξ^2 + 3.ξ' are functional signs. But in the case it must be laid down that the only thing 'ξ' does here is to show the places where the completing sign has to be inserted. It will be well not to employ this letter for any other purpose, and so, e.g., not instead of the '*x*' in our examples that serves to express generality.

It is a defect of the ordinary symbolism for differential quotients that in it the letter '*x*' has to serve both to show the places for the argument and to express generality, as in the equation:

$$\frac{d \cos \frac{x}{2}}{dx} = -\frac{1}{2} \sin \frac{x}{2}$$

From this there arises a difficulty. According to the general principles for the use of letters in arithmetic we should have to get a particular case of substituting a numeral for '*x*.' But the expression

$$\frac{d \cos \frac{2}{2}}{d2}$$

---

* In any case it is meant only for the exceptional case where we want to symbolize a function in isolation. In 'sin 2,' '*sin*' by itself already symbolizes the function.

is unintelligible, because we cannot recognize the function. We do not know whether it is

$$\cos\frac{(\ )}{2}, \text{ or } \cos\frac{2}{(\ )}, \text{ or } \cos\frac{(\ )}{(\ )}.$$

So we are forced to use the clumsy notation

$$\left(\frac{d\cos\frac{x}{2}}{dx}\right)x = 2$$

But the greater disadvantage is that it is thus made more difficult to see the nature of the function.

The peculiarity of functional signs, which we here called 'unsaturatedness,' naturally has something answering to it in the functions themselves. They too may be called 'unsaturated,' and in this way we mark them out as fundamentally different from numbers. Of course this is no definition; but likewise none is here possible.* I must confine myself to hinting at what I have in mind by means of a metaphorical expression, and here I rely on my reader's agreeing to meet me half-way.

If a function is completed by a number so as to yield a number, the second is called the value of the function for the first as argument. People have got used to reading the equation '$y = f(x)$' as 'y is a function of x.' There are two mistakes here: first, rendering the *equals*-sign as a copula; secondly, confusing the function with its value for an argument. From these mistakes has arisen the opinion that the function is a number, although a variable or indefinite one. We have seen, on the contrary, that there are no such numbers at all, and that functions are fundamentally different from numbers.

The endeavour to be brief has introduced many inexact expressions into mathematical language, and these have reacted by obscuring thought and producing faulty definitions. Mathematics ought properly to be a model of logical clarity. In actual fact there are perhaps no scientific works where you will find more wrong expressions, and consequently wrong thoughts, than in mathematical ones. Logical correctness should never be sacrificed to brevity of expression. It is therefore highly important to devise a mathematical language that combines the most rigorous accuracy with the greatest possible brevity. To this end a symbolic language would be best adapted, by means of which we could directly express thoughts in written or printed symbols without the intervention of spoken language.

---

* H. Hankel's definition, in his *untersuchungen über die unendlich oft oszillirenden und unstetigen Funktionen* (Universitatsprogramm. Tubingen 1870), §1, is useless, because of a vicious circle; it contains the expression '$f(x)$,' and this makes his definition presuppose the thing that is to be defined.

9
THE THOUGHT: A LOGICAL INQUIRY

The word "true" indicates the aim of logic as does "beautiful" that of aesthetics or "good" that of ethics. All sciences have truth as their goal; but logic is also concerned with it in a quite different way from this. It has much the same relation to truth as physics has to weight or heat. To discover truths is the task of all sciences; it falls to logic to discern the laws of truth. The word "law" is used in two senses. When we speak of laws of morals or the state we mean regulations which ought to be obeyed but with which actual happenings are not always in conformity. Laws of nature are the generalization of natural occurrences with which the occurrences are always in accordance. It is rather in this sense that I speak of laws of truth. This is, to be sure, not a matter of what happens so much as of what is. Rules for asserting, thinking, judging, inferring, follow from the laws of truth. And thus one can very well speak of laws of thought too. But there is an imminent danger here of mixing different things up. Perhaps the expression "law of thought" is interpreted by analogy with "law of nature" and the generalization of thinking as a mental occurrence is meant by it. A law of thought in this sense would be a psychological law. And so one might come to believe that logic deals with the mental process of thinking and the psychological laws in accordance with which it takes place. This would be a misunderstanding of the task of logic, for truth has not been given the place which is its due here. Error and superstition have causes just as much as genuine knowledge. The assertion both of what is false and of what is true takes place in accordance with psychological laws. A derivation from these and an explanation of a mental process that terminates in an assertion can never take the place of a proof of what is asserted. Could not logical laws also have played a part in this mental process? I do not want to dispute this, but when it is a question of truth possibility is not enough. For it is also possible that something not logical played a part in the process and deflected it from the truth. We can only decide this after we have discerned the laws of truth; but then we will probably be able to do without the derivation and explanation of the mental process if it is important to us to decide whether the assertion in which the process terminates is justified. In order to avoid this misunderstanding and to prevent the blurring of the boundary between psychology and logic, I assign to logic the task of discovering the laws of truth, not of assertion or thought. The meaning of the word "true" is explained by the laws of truth.

But first I shall attempt to outline roughly what I want to call true in this connexion. In this way other uses of our word may be excluded. It is not to be used here in the sense of "genuine" or "veracious", nor, as it sometimes occurs in the treatment of questions of art, when, for example, truth in art is discussed, when truth is set up as the goal of art, when the truth of a work of art or true feeling is spoken of. The word "true" is put in front of another word in order to show that this word is to be understood in its proper, unadulterated sense. This use too lies off the path followed here; that kind of truth is meant whose recognition is the goal of science.

Grammatically the word "true" appears as an adjective. Hence the desire arises to delimit more closely the sphere in which truth can be affirmed, in which truth comes into the question at all. One finds truth affirmed of pictures, ideas, statements, and thoughts. It is striking that visible and audible things occur here alongside things which cannot be perceived with the senses. This hints that shifts of meaning have taken place. Indeed! Is a picture, then, as a mere visible and tangible thing, really true, and a stone, a leaf, not true? Obviously one would not call a picture true unless there were an intention behind it. A picture must represent something. Furthermore, an idea is not called true in itself but only with respect to an intention that it should correspond to something. It might be supposed from this that truth consists in the correspondence of a picture with what it depicts. Correspondence is a relation. This is contradicted, however, by the use of the word "true", which is not a relation-word and contains no reference to anything else to which something must correspond. If I do not know that a picture is meant to represent Cologne Cathedral then I do not know with what to compare the picture to decide on its truth. A correspondence, moreover, can only be perfect if the corresponding things coincide and are, therefore, not distinct things at all. It is said to be possible to establish the authenticity of a banknote by comparing it stereoscopically with an authentic one. But it would be ridiculous to try to compare a gold piece with a twenty-mark note stereoscopically. It would only be possible to compare an idea with a thing if the thing were an idea too. And then, if the first did correspond perfectly with the second, they would coincide. But this is not at all what is wanted when truth is defined as the correspondence of an idea with something real. For it is absolutely essential that the reality be distinct from the idea. But then there can be no complete correspondence, no complete truth. So nothing at all would be true; for what is only half true is untrue. Truth cannot tolerate a more or less. But yet? Can it not be laid down that truth exists when there is correspondence in a certain respect? But in which? For what would we then have to do to decide whether something were true? We should have to inquire whether it were true that an idea and a reality, perhaps, corresponded in the laid-down respect. And then we should be confronted by a question of the same kind and the game could begin again. So the attempt to explain truth as correspondence collapses. And every other attempt to define truth collapses too. For in a definition certain characteristics would have to be stated. And in application to any particular case the question would always arise whether it were true that the characteristics were present. So one goes round in a circle. Consequently, it is probable that the content of the word "true" is unique and indefinable.

When one ascribes truth to a picture one does not really want to ascribe a property which belongs to this picture altogether independently of other things, but one always has something quite different in mind and one wants to say that that picture corresponds in some way to this thing. "My idea corresponds to Cologne Cathedral" is a sentence and the question now arises of the truth of this sentence. So what is improperly called the truth of pictures and ideas is reduced to the truth of sentences. What does one call a sentence? A series of sounds; but only when it has a sense, by which is not meant that every series of sounds that has sense is a sentence. And when we call a sentence true we really mean its sense is. From which it follows that it is for the sense of a sentence that the question of truth arises in general. Now is the sense of a sentence an idea? In any case being true does not consist in the correspondence of this sense with something else, for otherwise the question of truth would reiterate itself to infinity.

Without wishing to give a definition, I call a thought something for which the question of truth arises. So I ascribe what is false to a thought just as much as what is true.[1] So I can say: the thought is the sense of the sentence without wishing to say as well that the sense of every sentence is a thought. The thought, in itself immaterial, clothes itself in the material garment of a sentence and thereby becomes comprehensible to us. We say a sentence expresses a thought.

A thought is something immaterial and everything material and perceptible is excluded from this sphere of that for which the question of truth arises. Truth is not a quality that corresponds with a particular kind of sense-impression. So it is sharply distinguished from the qualities which we denote by the words "red", "bitter", "lilac-smelling". But do we not see that the sun has risen and do we not then also see that this is true? That the sun has risen is not an object which emits rays that reach my eyes, it is not a visible thing like the sun itself. That the sun has risen is seen to be true on the basis of sense-impressions. But being true is not a material, perceptible property. For being magnetic is also recognized on the basis of sense-impressions of something, though this property corresponds as little as truth

---

1 In a similar way it has perhaps been said 'a judgment is something which is either true or false'. In fact I use the word 'thought' in approximately the sense which 'judgment' has in the writings of logicians. I hope it will become clear in what follows why I choose 'thought'. Such an explanation has been objected to on the ground that in it a distinction is drawn between true and false judgments which of all possible distinctions among judgments has perhaps the least significance. I cannot see that it is a logical deficiency that a distinction is given with the explanation. As far as significance is concerned, it should not by any means be judged as trifling if, as I have said, the word 'true' indicates the aim of logic.

with a particular kind of sense-impressions. So far these properties agree. However, we need sense-impressions in order to recognize a body as magnetic. On the other hand, when I find that it is true that I do not smell anything at this moment, I do not do so on the basis of sense-impressions.

It may nevertheless be thought that we cannot recognize a property of a thing without at the same time realizing the thought that this thing has this property to be true. So with every property of a thing is joined a property of a thought, namely, that of truth. It is also worthy of notice that the sentence "I smell the scent of violets" has just the same content as the sentence "it is true that I smell the scent of violets". So it seems, then, that nothing is added to the thought by my ascribing to it the property of truth. And yet is it not a great result when the scientist after much hesitation and careful inquiry, can finally say "what I supposed is true"? The meaning of the word "true" seems to be altogether unique. May we not be dealing here with something which cannot, in the ordinary sense, be called a quality at all? In spite of this doubt I want first to express myself in accordance with ordinary usage, as if truth were a quality, until something more to the point is found.

In order to work out more precisely what I want to call thought, I shall distinguish various kinds of sentences.[1] One does not want to deny sense to an imperative sentence, but this sense is not such that the question of truth could arise for it. Therefore I shall not call the sense of an imperative sentence a thought. Sentences expressing desires or requests are ruled out in the same way. Only those sentences in which we communicate or state something come into the question. But I do not count among these exclamations in which one vents one's feelings, groaning, sighing, laughing, unless it has been decided by some agreement that they are to communicate something. But how about interrogative sentences? In a word-question we utter an incomplete sentence which only obtains a true sense through the completion for which we ask. Word-questions are accordingly left out of consideration here. Sentence-questions are a different matter. We expect to hear "yes" or "no". The answer "yes" means the same as an indicative sentence, for in it the thought that was already completely contained in the interrogative sentence is laid down as true. So a sentence-question can be formed from every indicative sentence. An exclamation cannot be regarded as a communication on this account, since no corresponding sentence-question can be formed. An interrogative sentence and an indicative one contain the same thought; but the indicative contains something else as well, namely, the assertion. The interrogative sentence contains something more too, namely a request. Therefore two things must be distinguished in an indicative sentence: the content, which it has in common with the corresponding sentence-question, and the assertion. The former is the thought, or at least contains the

---

1 I am not using the word 'sentence' here in a purely grammatical sense where it also includes subordinate clauses. An isolated subordinate clause does not always have a sense about which the question of truth can arise, whereas the complex sentence to which it belongs has such a sense.

thought. So it is possible to express the thought without laying it down as true. Both are so closely joined in an indicative sentence that it is easy to overlook their separability. Consequently we may distinguish:

(1) the apprehension of a thought – thinking,
(2) the recognition of the truth of a thought – judgment,[1]
(3) the manifestation of this judgment – assertion

We perform the first act when we form a sentence-question. An advance in science usually takes place in this way, first a thought is apprehended, such as can perhaps be expressed in a sentence-question, and, after appropriate investigations, this thought is finally recognized to be true. We declare the recognition of truth in the form of an indicative sentence. We do not have to use the word "true" for this. And even when we do use it the real assertive force lies, not in it, but in the form of the indicative sentence and where this loses its assertive force the word "true" cannot put it back again. This happens when we do not speak seriously. As stage thunder is only apparent thunder and a stage fight only an apparent fight, so stage assertion is only apparent assertion. It is only acting, only fancy. In his part the actor asserts nothing, nor does he lie, even if he says something of whose falsehood he is convinced. In poetry we have the case of thoughts being expressed without being actually put forward as true in spite of the form of the indicative sentence, although it may be suggested to the hearer to make an assenting judgment himself. Therefore it must still always be asked, about what is presented in the form of an indicative sentence, whether it really contains an assertion. And this question must be answered in the negative if the requisite seriousness is lacking. It is irrelevant whether the word "true" is used here. This explains why it is that nothing seems to be added to a thought by attributing to it the property of truth.

An indicative sentence often contains, as well as a thought and the assertion, a third component over which the assertion does not extend. This is often said to act on the feelings, the mood of the hearer or to arouse his imagination. Words like "alas" and "thank God" belong here. Such constituents of sentences are more noticeably prominent in poetry, but are seldom wholly absent from prose. They occur more rarely in mathematical, physical, or chemical than in historical expositions. What are called the humanities are more closely connected with poetry and are therefore less scientific than the exact sciences which are drier the more exact

---

[1] It seems to me that thought and judgment have not hitherto been adequately distinguished. Perhaps language is misleading. For we have no particular clause in the indicative sentence which corresponds to the assertion, that something is being asserted lies rather in the form of the indicative. We have the advantage in German that main and subordinate clauses are distinguished by the word-order. In this connexion it is noticeable that a subordinate clause can also contain an assertion and that often neither main nor subordinate clause express a complete thought by themselves but only the complex sentence does.

they are, for exact science is directed toward truth and only the truth. Therefore all constituents of sentences to which the assertive force does not reach do not belong to scientific exposition but they are sometimes hard to avoid, even for one who sees the danger connected with them. Where the main thing is to approach what cannot be grasped in thought by means of guesswork these components have their justification. The more exactly scientific an exposition is the less will the nationality of its author be discernible and the easier will it be to translate. On the other hand, the constituents of the language, to which I want to call attention here, make the translation of poetry very difficult, even make a complete translation almost always impossible, for it is in precisely that in which poetic value largely consists that language differ most.

It makes no difference to the thought whether I use the word "horse" or "steed" or "cart-horse" or "mare". The assertive force does not extend over that in which these words differ. What is called mood, fragrance, illumination in a poem, what is portrayed by cadence and rhythm, does not belong to the thought.

Much of language serves the purpose of aiding the hearer's understanding, for instance the stressing of part of a sentence by accentuation or word-order. One should remember words like "still" and "already" too. With the sentence "Alfred has still not come" one really says "Alfred has not come" and, at the same time, hints that his arrival is expected, but it is only hinted. It cannot be said that, since Alfred's arrival is not expected, the sense of the sentence is therefore false. The word "but" differs from "and" in that with it one intimates that what follows is in contrast with what would be expected from what preceded it. Such suggestions in speech make no difference to the thought. A sentence can be transformed by changing the verb from active to passive and making the object the subject at the same time. In the same way the dative may be changed into the nominative while "give" is replaced by "receive". Naturally such transformations are not indifferent in every respect; but they do not touch the thought, they do not touch what is true or false. If the inadmissibility of such transformations were generally admitted then all deeper logical investigation would be hindered. It is just as important to neglect distinctions that do not touch the heart of the matter as to make distinctions which concern what is essential. But what is essential depends on one's purpose. To a mind concerned with what is beautiful in language what is indifferent to the logician can appear as just what is important.

Thus the contents of a sentence often go beyond the thoughts expressed by it. But the opposite often happens too, that the mere wording, which can be grasped by writing or the gramophone does not suffice for the expression of the thought. The present tense is used in two ways: first, in order to give a date, second, in order to eliminate any temporal restriction where timelessness or eternity is part of the thought. Think, for instance, of the laws of mathematics. Which of the two cases occurs is not expressed but must be guessed. If a time indication is needed by the present tense one must know when the sentence was uttered to apprehend the thought correctly. Therefore the time of utterance is part of the expression of the thought. If someone wants to say the same today as he expressed yesterday using the word "today", he must replace this word with "yesterday". Although the thought is the same its verbal expression must be different so that the sense, which would otherwise be affected by the differing times of utterance, is readjusted. The case is the same with words like "here" and "there". In all such cases the mere wording, as it is given in writing, is not the complete expression of the thought, but the knowledge of certain accompanying conditions of utterance, which are used as means of expressing the thought, are needed for its correct apprehension. The pointing of fingers, hand movements, glances may belong here too. The same utterance containing the word "I" will express different thought in the mouths of different men, of which some may be true, others false.

The occurrence of the word "I" in a sentence gives rise to some questions.

Consider the following case. Dr. Gustav Lauben says, "I have been wounded". Leo Peters hears this and remarks some days later, "Dr. Gustav Lauben has been wounded". Does this sentence express the same thought as the one Dr. Lauben uttered himself? Suppose that Rudolph Lingens were present when Dr. Lauben spoke and now hears what is related by Leo Peter. If the same thought is uttered by Dr. Lauben and Leo Peter then Rudolph Lingens, who is fully master of the language and remembers what Dr. Lauben has said in his presence, must now know at once from Leo Peter's report that the same thing is under discussion. But knowledge of the language is a separate thing when it is a matter of proper names. It may well be the case that only a few people associate a particular thought with the sentence "Dr. Lauben has been wounded". In this case one needs for complete understanding a knowledge of the expression "Dr. Lauben". Now if both Leo Peter and Rudolph Lingens understand by "Dr. Lauben" the doctor who lives as the only doctor in a house known to both of them, then they both understand the sentence "Dr. Gustav Lauben has been wounded" in the same way, they associated the same thought with it. But it is also possible that Rudolph Lingens does not know Dr. Lauben personally and does not know that he is the very Dr. Lauben who recently said "I have been wounded". In this case Rudolph Lingens cannot know that the same thing is in question. I say, therefore, in this case: the thought which Leo Peter expresses is not the same as that which Dr. Lauben uttered.

Suppose further that Herbert Garner knows that Dr. Gustav Lauben was born on 13th September, 1875 in N.N. and this is not true of anyone else; against this, suppose that he does not know where Dr. Lauben now lives nor indeed anything about him. On the other hand, suppose Leo Peter does not know that Dr. Lauben was born on 13th December 1875, in N.N. Then as far as the proper name "Dr. Gustav Lauben" is concerned, Herbert Garner and Leo Peter do not speak the same language, since, although they do in fact refer to the same man with this name, they do not know that they do so. Therefore Herbert Garner does not associate the same thought with the sentence "Dr. Gustav Lauben has been wounded" as Leo Peter wants to express it. To avoid the drawback of Herbert Garner's and Leo Peter's not speaking the same language, I am assuming that Leo Peter uses the proper name

"Dr. Lauben" and Herbert Garner, on the other hand, uses the proper name "Gustav Lauben". Now it is possible that Herbert Garner takes the sense of the sentence "Dr. Lauben has been wounded" to be true while, misled by false information, taking the sense of the sentence "Gustav Lauben has been wounded" to be false. Under the assumptions given these thoughts are therefore different.

Accordingly, with a proper name, it depends on how whatever it refers to is presented. This can happen in different ways and every such way corresponds with a particular sense of a sentence containing an proper name. The different thoughts which thus result from the same sentence correspond in their truth-value, of course; that is to say, if one is true then all are true, and if one is false then all are false. Nevertheless their distinctness must be recognized. So it must really be demanded that a single way in which whatever is referred to is presented be associated with every proper name. It is often unimportant that this demand should be fulfilled but not always.

Now everyone is presented to himself in a particular and primitive way, in which he is presented to no-one else. So, when Dr. Lauben thinks that he has been wounded, he will probably take as a basis this primitive way in which he is presented to himself. And only Dr. Lauben himself can grasp thoughts determined in this way. But now he may want to communicate with others. He cannot communicate a thought which he alone can grasp. Therefore, if he now says "I have been wounded", he must use the "I" in a sense which can be grasped by others, perhaps in the sense of "he who is speaking to you at this moment", by doing which he makes the associated conditions of his utterance serve for the expression of his thought.[1]

Yet there is a doubt. Is it at all the same thought which first that man expresses and now this one?

A person who is still untouched by philosophy knows first of all things which he can see and touch, in short, perceive with the senses, such as trees, stones and houses, and he is convinced that another person equally can see and touch the same tree and the same stone which he himself sees and touches. Obviously a thought is not one of these things. Now can it, nevertheless, stand in the same relation to a person as to a tree?

Even an unphilosophical person soon finds it necessary to recognize an inner world distinct from the outer world, a world of sense-impressions, of creations of

---

[1] I am not in the happy position here of a mineralogist who shows his hearers a mountain crystal. I cannot put a thought in the hands of my readers with the request that they should minutely examine it from all sides. I have to content myself with presenting the reader with a thought, in itself immaterial, dressed in sensible linguistic form. The metaphorical aspect of language presents difficulties. The sensible always breaks in and makes expression metaphorical and so improper. So a battle with language takes place and I am compelled to occupy myself with language although it is not my proper concern here. I hope I have succeeded in making clear to my readers what I want to call a thought.

his imagination, of sensations, of feelings and moods, a world of inclinations, wishes and decisions. For brevity I want to collect all these, with the exception of decisions, under the word "idea".

Now do thoughts belong to this inner world? Are they ideas? They are obviously not decisions. How are ideas distinct from the things of the outer world? First:

Ideas cannot be seen or touched, cannot be smelled, nor tasted, nor heard.

Secondly: ideas are had. One has sensations, feelings, moods, inclinations, wishes. An idea which someone has belongs to the content of his consciousness.

The field and the frogs in it, the sun which shines on them are there no matter whether I look at them or not, but the sense-impression I have of green exists only because of me, I am its bearer. It seems absurd to us that a pain, a mood, a wish should rove about the world without a bearer, independently. An experience is impossible without an experient. The inner world presupposes the person whose inner world it is.

Thirdly: ideas need a bearer. Things of the outer world are however independent.

My companion and I are convinced that we both see the same field; but each of us has a particular sense-impression of green. I notice a strawberry among the green strawberry leaves. My companion does not notice it, he is colour-blind. The colour-impression, which he receives from the strawberry, is not noticeably different from the one he receives from the leaf. Now does my companion see the green leaf as red, or does he see the red berry as green, or does he see both as of one colour with which I am not acquainted at all? These are unanswerable, indeed really nonsensical, questions. For when the word "red" does not state a property of things but is supposed to characterize sense-impressions belonging to my consciousness, it is only applicable within the sphere of my consciousness. For it is impossible to compare my sense-impression with that of someone else. For that it would be necessary to bring together in one consciousness a sense-impression, belonging to one consciousness, with a sense-impression belonging to another consciousness. Now even if it were possible to make an idea disappear from one consciousness and, at the same time, to make an idea appear in another consciousness, the question whether it were the same idea in both would still remain unanswerable. It is so much of the essence of each of my ideas to be the content of my consciousness, that every idea of another person is, just as such, distinct from mine. But might it not be possible that my ideas, the entire content of my consciousness might be at the same time the content of a more embracing, perhaps divine, consciousness? Only if I were myself part of the divine consciousness. But then would they really be my ideas, would I be their bearer? This oversteps the limits of human understanding to such an extent that one must leave its possibility out of account. In any case it is impossible for us as men to compare another person's ideas with our own. I pick the strawberry, I hold it between my fingers. Now my companion sees it too, this very same strawberry; but each of us has his own idea. No other person has my pain. Someone can have sympathy for me but still my pain always belongs to me and his sympathy to him. He does not have my pain and I do not have his sympathy.

Fourthly: every idea has only one bearer; no two men have the same idea.

For otherwise it would exist independently of this person and independently of that one. Is that lime-tree my idea? By using the expression "that lime-tree" in this question I have really already anticipated the answer, for with this expression I want to refer to what I see and to what other people can also look at and touch. There are now two possibilities. If my intention is realized when I refer to something with the expression "that lime-tree" then the thought expressed in the sentence "that lime-tree is my idea" must obviously be negated. But if my intention is not realized, if I only think I see without really seeing, if on that account the designation "that lime-tree" is empty, then I have gone astray into the sphere of fiction without knowing it or wanting to. In that case neither the content of the sentence "that lime-tree is my idea" nor the content of the sentence "that lime-tree is not my idea" is true, for in both cases I have a statement which lacks an object. So then one can only refuse to answer the question for the reason that the content of the sentence "that lime-tree is my idea" is a piece of fiction. I have, naturally, got an idea then, but I am not referring to this with the words "that lime-tree". Now someone may really want to refer to one of his ideas with the words "that lime-tree". He would then be the bearer of that to which he wants to refer with those words, but then he would not see that lime-tree and no-one else would see it or be its bearer.

I now return to the question: is a thought an idea? If the thought I express in the Pythagorean theorem can be recognized by others just as much as by me then it does not belong to the content of my consciousness, I am not its bearer; yet I can, nevertheless, recognize it to be true. However, if it is not the same thought at all which is taken to be the content of the Pythagorean theorem by me and by another person, one should not really say "the Pythagorean theorem" but "my Pythagorean theorem", "his Pythagorean theorem" and these would be different; for the sense belongs necessarily to the sentence. Then my thought can be the content of my consciousness and his thought the content of his. Could the sense of my Pythagorean theorem be true while that of his was false? I said that the word "red" was applicable only in the sphere of my consciousness if it did not state a property of things but was supposed to characterize one of my sense-impressions. Therefore the words "true" and "false", as I understand them, could also be applicable only in the sphere of my consciousness, if they were not supposed to be concerned with something of which I was not the bearer, but were somehow appointed to characterize the content of my consciousness. Then the truth would be restricted to the content of my consciousness and it would remain doubtful whether anything at all comparable occurred in the consciousness of others.

If every thought requires a bearer, to the contents of whose consciousness it belongs, then it would be a thought of this bearer only and there would be no science common to many, on which many could work. But I, perhaps, have my science, namely, a whole of thought whose bearer I am and another person has his. Each of us occupies himself with the contents of his own consciousness. No contradiction between the two sciences would then be possible and it would really be idle to dispute about truth, as idle, indeed almost ludicrous, as it would be for two people to dispute whether a hundred-mark note were genuine, where each meant the one he himself had in his pocket and understood the word "genuine" in his own particular sense. If someone takes thoughts to be ideas, what he then recognizes to be true is, on his own view, the content of his consciousness and does not properly concern other people at all. If he were to hear from me the opinion that a thought is not an idea he could not dispute it, for, indeed, it would not now concern him.

So the results seem to be: thoughts are neither things of the outer world nor ideas.

A third realm must be recognized. What belongs to this corresponds with ideas, in that it cannot be perceived by the senses, but with things, in that it needs no bearer to the contents of whose consciousness to belong. Thus the thought, for example, which we expressed in the Pythagorean theorem is timelessly true, true independently of whether anyone takes it to be true. It needs no bearer. It is not true for the first time when it is discovered, but is like a planet which, already before any has seen it, has been in interaction with other planets.[1]

But I think I hear an unusual objection. I have assumed several times that the same thing that I see can also be observed by other people. But how could this be the case, if everything were only a dream? If I only dreamed I was walking in the company of another person, if I only dreamed that my companion saw the green field as I did, if it were all only a play performed on the stage of my consciousness, it would be doubtful whether there were things of the outer world at all. Perhaps the realm of things is empty and I see no things and no men, but have only ideas of which I myself am the bearer. An idea, being something which can as little exist independently of me as my feeling of fatigue, cannot be a man, cannot look at the same field together with me, cannot see the strawberry I am holding. It is quite incredible that I should really have only my inner world instead of the whole environment, in which I am supposed to move and to act. And yet it is an inevitable consequence of the thesis that only what is my idea can be the object of my awareness. What would follow from this thesis if it were true? Would there then be other men? It would certainly be possible but I should know nothing of it. For a man cannot be my idea, consequently, if our thesis were true, he also cannot be an object of my awareness. And so the ground would be removed from

---

[1] One sees a thing, one has an idea, one apprehends or thinks a thought. When one apprehends or thinks a thought one does not create it but only comes to stand in a certain relation, which is different from seeing a thing or having an idea, to what already existed beforehand.

under any process of thought in which I might assume that something was an object for another person as for myself, for even if this were to happen I should know nothing of it. It would be impossible for me to distinguish that of which I was the bearer from that of which I was not. In judging something not to be my idea I would make it the object of my thinking and, therefore, my idea. On this view, is there a green field? Perhaps, but it would not be visible to me. For if a field is not my idea, it cannot, according to our thesis, be an object of my awareness. But if it is my idea it is invisible, for ideas are not visible. I can indeed have the idea of a green field, but this is not green for there are no green ideas. Does a shell weighing a hundred kilogrammes exist, according to this view? Perhaps, but I could know nothing of it. If a shell is not my idea then, according to our thesis, it cannot be an object of my awareness, of my thinking. But if a shell were my idea, it would have no weight. I can have an idea of a heavy shell. This then contains the idea of weight as a part-idea. But this part-idea is not a property of the whole idea any more than Germany is a property of Europe. So it follows:

Either the thesis that only what is my idea can be the object of my awareness is false, or all my knowledge and perception is limited to the range of my ideas, to the stage of my consciousness. In this case I should have only an inner world and I should know nothing of other people.

It is strange how, upon such reflections, the opposites collapse into each other. There is, let us suppose, a physiologist of the senses. As is proper for a scholarly scientist, he is, first of all, far from supposing the things he is convinced he sees and touches to be his ideas. On the contrary, he believes that in sense-impressions he has the surest proof of things which are wholly independent of his feeling, imagining, thinking, which have no need of his consciousness. So little does he consider nerve-fibres and ganglion-cells to be the content of his consciousness that he is, on the contrary, rather inclined to regard his consciousness as dependent on nerve-fibres and ganglion-cells. He establishes that light-rays, refracted in the eye, strike the visual nerve-endings and bring about a change, a stimulus, there. Some of it is transmitted through nerve-fibres and ganglion-cells. Further processes in the nervous system are perhaps involved, colour-impressions arise and these perhaps join themselves to what we call the idea of a tree. Physical, chemical and physio-logical occurrences insert themselves between the tree and my idea. These are immediately connected with my consciousness but, so it seems, are only occurrences in my nervous system and every spectator of the tree has his particular occurrences in his particular nervous system. Now the light-rays, before they enter my eye, may be reflected by a mirror and be spread further as if they came from a place behind the mirror. The effects on the visual nerves and all that follows will now take place just as they would if the light-rays had come from a tree behind the mirror and had been transmitted undisturbed to the eye. So an idea of a tree will finally occur even though such a tree does not exist at all. An idea, to which nothing at all corresponds, can also arise through the bending of light, with the mediation of the eye and the nervous system. But the stimulation of the visual nerves need not even happen through light. If lightning strikes near us we believe we see flames, even though we cannot see the lightening itself. In this case the visual nerve is perhaps stimulated by electric currents which originate in our body in consequence of the flash of lightening. If the visual nerve is stimulated by this means, just as it would be stimulated by light-rays coming from flames, then we believe we see flames. It just depends on the stimulation of the visual nerve, it is indifferent how that itself comes about.

One can go a step further still. This stimulation of the visual nerve is not actually immediately given, but is only a hypothesis. We believe that a thing, independent of us, stimulates a nerve and by this means produces a sense-impression, but, strictly speaking, we experience only the end of this process which projects into our consciousness. Could not this sense-impression, this sensation, which we attribute to a nerve-stimulation, have other causes also, as the same nerve-stimulation can arise in different ways? If we call what happens in our consciousness idea, then we really experience only ideas but not their causes. And if the scientist wants to avoid all mere hypothesis, then only ideas are left for him, everything resolves into ideas, the light-rays, nerve-fibres and ganglion-cells from which he started. So he finally undermines the foundations of his own construction. Is everything an idea? Does everything need a bearer, without which it could have no stability? I have considered myself as the bearer of my ideas, but am I not an idea myself? It seems to me as if I were lying in a deck-chair, as if I could see the toes of a pair of waxed boots, the front part of a pair of trousers, a waistcoat, buttons, part of a jacket, in particular sleeves, two hands, the hair of a beard, the blurred outline of a nose. Am I myself this entire association of visual impressions, this total idea? It also seems to me as if I see a chair over there. It is an idea. I am not actually much different from this myself, for am I not myself just an association of sense-impressions, an idea? But where then is the bearer of these ideas? How do I come to single out one of these ideas and set it up as the bearer of the rest? Why must it be the idea which I choose to call "I"? Could I not just as well choose the one that I am tempted to call a chair? Why, after all, have a bearer for ideas at all? But this would always be something independent, needing no extraneous bearer. If everything is idea, then there is no bearer of ideas. And so now, once again, I experience a change into the opposite. If there is no bearer of ideas then there are also no ideas, for ideas need a bearer without which they cannot exist. If there is no ruler, there are also no subjects. The dependence, which I found myself induced to confer on the experience as opposed to the experient, is abolished if there is no more bearer. What I called ideas are then independent objects. Every reason is wanting for granting an exceptional position to that object which I call "I".

But is that possible? Can there be an experience without someone to experience it? What would this whole play be without an onlooker? Can there be a pain without someone who has it? Being experienced is necessarily connected with pain, and someone experiencing is necessarily connected with being experienced. But there is something which is not my idea and yet which can be the object of my awareness, of my thinking I am myself of this nature. Or can I be part of the content of my consciousness while another part is, perhaps, an idea of the moon? Does this perhaps take place when I judge that I am looking at the moon? Then this first part would have a consciousness and part of the content of this consciousness would be I myself once more. And so on. Yet it is surely inconceivable that I should be boxed into myself in this way to infinity, for then there would not be only one I but infinitely many. I am not my own idea and if I assert something about myself, *e.g.* that I do not feel any pain at this moment, then my judgment concerns something which is not a content of my consciousness, is not my idea, that is me myself. Therefore that about which I state something is not necessarily my idea. But, someone perhaps objects, if I think I have no pain at the moment, does not the word 'I' nevertheless correspond with something in the content of my consciousness and is that not an idea? That may be. A certain idea in my consciousness may be associated with the idea of the word 'I'. But then it is an idea among other ideas and I am its bearer as I am the bearer of the other ideas. I have an idea of myself but I am not identical with this idea. What is a content of my consciousness, my idea, should be sharply distinguished from what is an object of my thought. Therefore the thesis that only what belongs to the content of my consciousness can be the object of my awareness, of my thought, is false.

Now the way is clear for me to recognize another person as well as to be an independent bearer of ideas. I have an idea of him but I do not confuse it with him himself. And if I state something about my brother I do not state it about the idea that I have of my brother.

The invalid who has a pain is the bearer of this pain, but the doctor in attendance who reflects on the cause of this pain is not the bearer of the pain. He does not imagine he can relieve the pain by anaesthetizing himself. An idea in the doctor's mind may very well correspond to the pain of the invalid but that is not the pain and not what the doctor is trying to remove. The doctor might consult another doctor. Then one must distinguish: first, the pain whose bearer is the invalid, second, the first doctor's idea of this pain, third, the second doctor's idea of this pain. This idea does indeed belong to the content of the second doctor's consciousness, but it is not the object of his reflection, it is rather an aid to reflection, as a drawing can be such an aid perhaps. Both doctors have the invalid's pain, which they do not bear, as their common object of thought. It can be seen from this that not only a thing but also an idea can be the common object of thought of people who do not have the idea.

So, it seems to me, the matter becomes intelligible. If man could not think and could not take something of which he was not the bearer as the object of his thought he would have an inner world but no outer world. But may this not be based on a mistake? I am convinced that the idea I associate with the words 'my brother' corresponds to something that is not my idea and about which I can say something. But may I not be making a mistake about this? Such mistakes do happen. We then, against our will, lapse into fiction. Indeed! By the step with which I secure an environment for myself I expose myself to the risk of error. And here I come up against a further distinction between my inner and outer worlds. I cannot doubt that I have a visual impression of green but it is not so certain that I see a lime-leaf. So, contrary to widespread views, we find certainty in the inner world while doubt never altogether leaves us in our excursions into the outer world. It is difficult in many cases, nevertheless, to distinguish probability from certainty here, so we can presume to judge about things in the outer world. And we must presume this even at the risk of error if we do not want to succumb to far greater dangers.

In consequence of these last considerations I lay down the following: not everything that can be the object of my understanding is an idea. I, as a bearer of ideas, am not myself an idea. Nothing now stands in the way of recognizing other people to be bearers of ideas as I am myself. And, once given the possibility, the probability is very great, so great that it is in my opinion no longer distinguishable from certainty. Would there be a science of history otherwise? Would not every precept of duty, every law otherwise come to nothing? What would be left of religion? The natural sciences too could only be assessed as fables like astrology and alchemy. Thus the reflections I have carried on, assuming that there are other people besides myself who can take the same thing as the object of their consideration, of their thinking, remain essentially unimpaired in force.

Not everything is an idea. Thus I can also recognize the thought, which other people can grasp just as much as I, as being independent of me. I can recognize a science in which many people can be engaged in research. We are not bearers of thoughts as we are bearers of our ideas. We do not have a thought as we have, say, a sense-impression, but we also do not see a thought as we see, say, a star. So it is advisable to choose a special expression and the word 'apprehend' offers itself for the purpose. A particular mental capacity, the power of thought, must correspond to the apprehension[1] of thought. In thinking we do not produce thoughts but we apprehend them. For what I have called thought stands in the closest relation to truth. What I recognize as true I judge to be true quite independently of my recognition of its truth and of my thinking about it. That someone thinks it has nothing to do with the truth of a thought. 'Facts, facts, facts' cries the scientist if he wants to emphasise the necessity of a firm foundation for science. What is a fact? A fact is a thought that is true. But the scientist will surely not recognise something which depends on men's varying states of mind to be the firm foundation of science. The work of science does not consist of creation but of the discovery of true thoughts. The astronomer can apply a mathematical truth in the investigation of long past events which took place when on earth at least no one had yet recognized that

---

1 The expression 'apprehend' is as metaphorical as 'content of consciousness'. The nature of language does not permit anything else. What I hold in my hand can certainly be regarded as the content of my hand but is all the same content of my hand in quite a different way from the bones and muscles of which it is made and their tensions, and is much more extraneous to it than they are.

truth. He can do this because the truth of a thought is timeless. Therefore that truth cannot have come into existence with its discovery.

Not everything is an idea. Otherwise psychology would contain all the sciences within it or at least it would be the highest judge over all the sciences. Otherwise psychology would rule over logic and mathematics. But nothing would be a greater misunderstanding of mathematics than its subordination to psychology. Neither logic nor mathematics has the task of investigating minds and the contents of consciousness whose bearer is a single person. Perhaps their task could be represented rather as the investigation of the mind, of the mind not of minds.

The apprehension of a thought presupposes someone who apprehends it, who thinks. He is the bearer of the thinking but not of the thought. Although the thought does not belong to the contents of the thinker's consciousness yet something in his consciousness must be aimed at the thought. But this should not be confused with the thought itself. Similarly Algol itself is different from the idea someone has of Algol.

The thought belongs neither to my inner world as an idea nor yet to the outer world of material, perceptible things.

This consequence, however cogently it may follow from the exposition, will nevertheless not perhaps be accepted without opposition. It will, I think, seem impossible to some people to obtain information about something not belonging to the inner world except by sense-perception. Sense-perception indeed is often thought to be the most certain, even to be the sole, source of knowledge about everything that does not belong to the inner world. But with what right? For sense-impressions are necessary constituents of sense-perceptions and are a part of the inner world. In any case two men do not have the same, though they may have similar, sense-impressions. These alone do not disclose the outer world to us. Perhaps there is a being that has only sense-impressions without seeing or touching things. To have visual impressions is not to see things. How does it happen that I see the tree just there where I do see it? Obviously it depends on the visual impressions I have and on the particular type which occur because I see with two eyes. A particular image arises, physically speaking, on each of the two retinas. Another person sees the tree in the same place. He also has two retinal images but they differ from mine. We must assume that these retinal images correspond to our impressions. Consequently we have visual impressions, not only not the same, but markedly different from each other. And yet we move about in the same outer world. Having visual impressions is certainly necessary for seeing things but not sufficient. What must still be added is non-sensible. And yet this is just what opens up the outer world for us; for without this non-sensible something everyone would remain shut up in his inner world. So since the answer lies in the non-sensible, perhaps something non-sensible could also lead us out of the inner world and enable us to grasp thoughts where no sense-impressions were involved. Outside one's inner world one would have to distinguish the proper outer world of sensible, perceptible things from the realm of the nonsensibly perceptible. We should need something non-sensible for the recognition of both realms but for the sensible perception of things we should need sense-impressions as well and these belong entirely to the inner world. So that in which the distinction between the way in which a thing and a thought is given mainly consists is something which is attributable, not to both realms, but to the inner world. Thus I cannot find this distinction to be so great that on its account it would be impossible for a thought to be given that did not belong to the inner world.

The thought, admittedly, is not something which it is usual to call real. The world of the real is a world in which this acts on that, changes it and again experiences reactions itself and is changed by them. All this is a process in time. We will hardly recognize what is timeless and unchangeable as real. Now is the thought changeable or is it timeless? The thought we express by the Pythagorean theorem is surely timeless, eternal, unchangeable. But are there not thoughts which are true today but false in six months time? The thought, for example, that the tree there is covered with green leaves, will surely be false in six months time. No, for it is not the same thought at all. The words 'this tree is covered with green leaves' are not sufficient by themselves for the utterance, the time of utterance is involved a well. Without the time-indication this gives we have no complete thought, *i.e.* no thought at all. Only a sentence supplemented by a time-indication and complete in every respect expresses a thought. But this, if it is true, is true not only today or tomorrow but timelessly. Thus the present tense in 'is true' does not refer to the speaker's present but is, if the expression be permitted, a tense of timelessness. If we use the mere form of the indicative sentence, avoiding the word 'true', two things must be distinguished, the expression of the thought and the assertion. The time-indication that may be contained in the sentence belongs only to the expression of the thought, while the truth, whose recognition lies in the form of the indicative sentence, is timeless. Yet the same words, on account of the variability of language with time, take on another sense, express another thought; this change, however, concerns only the linguistic aspect of the matter.

And yet! What value could there be for us in the eternally unchangeable which could neither undergo effects nor have effect on us? Something entirely and in every respect inactive would be unreal and non-existent for us. Even the timeless, if it is to be anything for us, must somehow be implicated with the temporal. What would a thought be for me that was never apprehended by me? But by apprehending a thought I come into a relation to it and it to me. It is possible that the same thought that is thought by me today was not thought by me yesterday. In this way the strict timelessness is of course annulled. But one is inclined to distinguish between essential and inessential properties and to regard something as timeless if the changes it undergoes involve only its inessential properties. A property of a thought will be called inessential which consists in, or follows from the fact that, it is apprehended by a thinker.

How does a thought act? By being apprehended and taken to be true. This is a process in the inner world of a thinker which can have further consequences in this inner world and which, encroaching on the sphere of the will, can also make itself noticeable in the outer world. If, for example, I grasp the thought which we express by the theorem of Pythagoras, the consequence may be that I recognize it to be true and, further, that I apply it, making a decision which brings about the acceleration of masses. Thus our actions are usually prepared by thinking and judgment. And so thought can have an indirect influence on the motion of masses. The influence of one person on another is brought about for the most part by thoughts. One communicates a thought. How does this happen? One brings about changes in the common outside world which, perceived by another person, are supposed to induce him to apprehend a thought and take it to be true. Could the great events of world history have come about without the communication of thoughts? And yet we are inclined to regard thoughts as unreal because they appear to be without influence on events, while thinking, judging, stating, understanding and the like are facts of human life. How much more real a hammer appears compared with a thought. How different the process of handing over a hammer is from the communication of a thought. The hammer passes from one control to another, it is gripped, it undergoes pressure and on account of this its density, the disposition of its parts, is changed in places. There is nothing of all this with a thought. It does not leave the control of the communicator by being communicated, for after all a person has no control over it. When a thought is apprehended, it at first only brings about changes in the inner world of the apprehender, yet it remains untouched in its true essence, since the changes it undergoes involve only inessential properties. There is lacking here something we observe throughout the order of nature: reciprocal action. Thoughts are by no means unreal but their reality is of quite a different kind from that of things. And their effect is brought about by an act of the thinker without which they would be ineffective, at least as far as we can see. And yet the thinker does not create them but must take them as they are. They can be true without being apprehended by a thinker and are not wholly unreal even then, at least if they could be apprehended and by this means be brought into operation.

WORKS OF
BERTRAND RUSSELL

10
MATHEMATICS AND THE METAPHYSICIANS

The nineteenth century, which prided itself upon the invention of steam and evolution, might have derived a more legitimate title to fame from the discovery of pure mathematics. This science, like most others, was baptised long before it was born; and thus we find writers before the nineteenth century alluding to what they called pure mathematics. But if they had been asked what this subject was, they would only have been able to say that it consisted of Arithmetic, Algebra, Geometry, and so on. As to what these studies had in common, and as to what distinguished them from applied mathematics, our ancestors were completely in the dark.

Pure mathematics was discovered by Boole, in a work which he called the *Laws of Thought* (1854). This work abounds in asseverations that it is not mathematical, the fact being that Boole was too modest to suppose his book the first ever written on mathematics. He was also mistaken in supposing that he was dealing with the laws of thought: the question how people actually think was quite irrelevant to him, and if his book had really contained the laws of thought, it was curious that no one should ever have thought in such a way before. His book was in fact concerned with formal logic, and this is the same thing as mathematics.

Pure mathematics consists entirely of assertions to the effect that, if such and such a proposition is true of *anything*, then such and such another proposition is true of that thing. It is essential not to discuss whether the first proposition is really true, and not to mention what the anything is, of which it is supposed to be true. Both these points belong to applied mathematics. We start, in pure mathematics, from certain rules of inference, by which we can infer that *if* one proposition is true, then so is some other proposition. These rules of inference constitute the major part of the principles of formal logic. We then take any hypothesis that seems amusing, and deduce its consequences. *If* our hypothesis is about *anything*, and not about some one or more particular things, then our deductions constitute mathematics. Thus mathematics may be defined as the subject in which we never know what we are talking about, nor whether what we are saying is true. People who have been puzzled by the beginnings of mathematics will, I hope, find comfort in this definition, and will probably agree that it is accurate.

As one of the chief triumphs of modern mathematics consists in having discovered what mathematics really is, a few more words on this subject may not be amiss. It is common to start any branch of mathematics – for instance, Geometry – with a certain number of primitive ideas, supposed incapable of definition, and a certain number of primitive propositions or axioms, supposed incapable of proof. Now the fact is that, though there are indefinables and indemonstrables in every branch of applied mathematics, there are none in pure mathematics except such as belong to general logic. Logic, broadly speaking, is distinguished by the fact that its propositions can be put into a form in which they apply to anything whatever. All pure mathematics – Arithmetic, Analysis, and Geometry – is built up by combinations of the primitive ideas of logic, and its propositions are deduced from the general axioms of logic, such as the syllogism and the other rules of inference. And this is no longer a dream or an aspiration. On the contrary, over the greater and more difficult part of the domain of mathematics, it has been already accomplished; in the few remaining cases, there is no special difficulty, and it is now being rapidly achieved. Philosophers have disputed for ages whether such deduction was possible; mathematicians have sat down and made the deduction. For the philosophers there is now nothing left but graceful acknowledgments.

The subject of formal logic, which has thus at last shown itself to be identical with mathematics, was, as every one knows, invented by Aristotle, and formed the chief study (other than theology) of the Middle Ages. But Aristotle never got beyond the syllogism, which is a very small part of the subject, and the schoolmen never got beyond Aristotle. If any proof were required of our superiority to the mediæval doctors, it might be found in this. Throughout the Middle Ages, almost all the best intellects devoted themselves to formal logic, whereas in the nineteenth century only an infinitesimal proportion of the world's thought went into this subject. Nevertheless, in each decade since 1850 more has been done to advance the subject than in the whole period from Aristotle to Leibniz. People have discovered how to make reasoning symbolic, as it is in Algebra, so that deductions are effected by mathematical rules. They have discovered many rules besides the syllogism, and a new branch of logic, called the Logic of Relatives,[1] has been invented to deal with topics that wholly surpassed the powers of the old logic, though they form the chief contents of mathematics.

It is not easy for the lay mind to realize the importance of symbolism in discussing the foundations of mathematics, and the explanation may perhaps seem strangely paradoxical. The fact is that symbolism is useful because it makes things difficult. (This is not true of the advanced parts of mathematics, but only of the beginnings.) What we wish to know is, what can be deduced from what. Now, in the beginnings, everything is self-evident; and it is very hard to see whether one self-evident proposition follows from another or not. Obviousness is always the

---

1   This subject is due in the main to Mr. C.S. Peirce.

enemy to correctness. Hence we invent some new and difficult symbolism, in which nothing seems obvious. Then we set up certain rules for operating on the symbols, and the whole thing becomes mechanical. In this way we find out what must be taken as premiss and what can be demonstrated or defined. For instance, the whole of Arithmetic and Algebra has been shown to require three indefinable notions and five indemonstrable propositions. But without a symbolism it would have been very hard to find this out. It is so obvious that two and two are four, that we can hardly make ourselves sufficiently sceptical to doubt whether it can be proved. And the same hold in other cases where self-evident things are to be proved.

But the proof of self-evident propositions may seem, to the uninitiated, a somewhat frivolous occupation. To this we might reply that it is often by no means self-evident that one obvious proposition follows from another obvious proposition; so that we are really discovering new truths when we prove what is evident by a method which is not evident. But a more interesting retort is, that since people have tried to prove obvious propositions, they have found that many of them are false. Self-evidence is often a mere will-o'-the-wisp, which is sure to lead us astray if we take it as our guide. For instance, nothing is plainer than that a whole always has more terms than a part, or that a number is increased by adding one to it. But these propositions are now known to be usually false. Most numbers are infinite, and if a number is infinite you may add ones to it as long as you like without disturbing it in the least. One of the merits of a proof is that it instils a certain doubt as to the result proved; and when what is obvious can be proved in some cases, but not in others, it becomes possible to suppose that in these other cases it is false.

The great master of the art of formal reasoning, among the men of our own day, is an Italian, Professor Peano, of the University of Turin.[1] He has reduced the greater part of mathematics (and he or his followers will, in time, have reduced the whole) to strict symbolic form, in which there are no words at all. In the ordinary mathematical books, there are no doubt fewer words than most readers would wish. Still, little phrases occur, such as *therefore, let us assume, consider*, or *hence it follows*. All these, however, are a concession, and are swept away by Professor Peano. For instance, if we wish to learn the whole of Arithmetic, Algebra, the Calculus, and indeed all that is usually called pure mathematics (except Geometry), we must start with a dictionary of three words. One symbol stands for *zero*, another for *number*, and a third for *next after*. What these ideas mean, it is necessary to know if you wish to become an arithmetician. But after symbols have been invented for these three ideas, not another word is required in the whole development. All future symbols are symbolically explained by means of these three. Even these three can be explained by means of the notions of *relation*, and *class*; but this

---

[1] I ought to have added Frege, but his writings were unknown to me when this article was written. [Note added in 1917.]

requires the Logic of Relations, which Professor Peano has never taken up. It must be admitted that what a mathematician has to know to begin with is not much. There are at most a dozen notions out of which all the notions in all pure mathematics (including Geometry) are compounded. Professor Peano, who is assisted by a very able school of young Italian disciples, has shown how this may be done; and although the method which he has invented is capable of being carried a good deal further than he has carried it, the honour of the pioneer must belong to him.

Two hundred years ago, Leibniz foresaw the science which Peano has perfected, and endeavoured to create it. He was prevented from succeeding by respect for the authority of Aristotle, whom he could not believe guilty of definite, formal fallacies; but the subject which he desired to create now exists, in spite of the patronising contempt with which his schemes have been treated by all superior persons. From this "Universal Characteristic," as he called it, he hoped for a solution of all problems, and an end to all disputes. "If controversies were to arise," he says, "there would be no more need of disputation between two philosophers than between two accountants. For it would suffice to take their pens in their hands, to sit down to their desks, and to say to each other (with a friend as witness, if they liked), 'Let us calculate.'" This optimism now has appeared to be somewhat excessive; there still are problems whose solution is doubtful, and disputes which calculation cannot decide. But over an enormous field of what was formerly controversial, Leibniz's dream has become sober fact. In the whole philosophy of mathematics, which used to be at least as full of doubt as any other part of philosophy, order and certainty have replaced the confusion and hesitation which formerly reigned. Philosophers, of course, have not yet discovered this fact, and continue to write on such subjects in the old way. But mathematicians, at least in Italy, have now the power of treating the principles of mathematics in an exact and masterly manner, by means of which the certainty of mathematics extends also to mathematical philosophy. Hence many of the topics which used to be placed among the great mysteries – for example, the natures of infinity, of continuity, of space, time and motion – are now no longer in any degree open to doubt or discussion. Those who wish to know the nature of these things need only read the works of such men as Peano or Georg Cantor; they will there find exact and indubitable expositions of all these quondam mysteries.

In this capricious world, nothing is more capricious than posthumous fame. One of the most notable examples of posterity's lack of judgment is the Eleatic Zeno. This man, who may be regarded as the founder of the philosophy of infinity, appears in Plato's Parmenides in the privileged position of instructor to Socrates. He invented four arguments, all immeasurably subtle and profound, to prove that motion is impossible, that Achilles can never overtake the tortoise, and that an arrow in flight is really at rest. After being refuted by Aristotle, and by every subsequent philosopher from that day to our own, these arguments were reinstated, and made the basis of a mathematical renaissance, by a German professor, who probably never dreamed of any connection between himself and Zeno.

Weierstrass,[1] by strictly banishing from mathematics the use of infinitesimals, has at last shown that we live in an unchanging world, and that the arrow in its flight is truly at rest. Zeno's only error lay in inferring (if he did infer) that, because there is no such thing as a state of change, therefore the world is in the same state at any one time as at any other. This is a consequence which by no means follows; and in this respect, the German mathematician is more constructive than the ingenious Greek. Weierstrass has been able, by embodying his views in mathematics, where familiarity with truth eliminates the vulgar prejudices of common sense, to invest Zeno's paradoxes with the respectable air of platitudes; and if the result is less delightful to the lover of reason than Zeno's bold defiance, it is at any rate more calculated to appease the mass of academic mankind.

Zeno was concerned, as a matter of fact, with three problems, each presented by motion, but each more abstract than motion, and capable of a purely arithmetical treatment. These are the problems of the infinitesimal, the infinite, and continuity. To state clearly the difficulties involved, was to accomplish perhaps the hardest part of the philosopher's task. This was done by Zeno. From him to our own day, the finest intellects of each generation in turn attacked the problems, but achieved, broadly speaking, nothing. In our own time, however, three men – Weierstrass, Dedekind, and Cantor – have not merely advanced the three problems, but have completely solved them. The solutions, for those acquainted with mathematics, are so clear as to leave no longer the slightest doubt or difficulty. This achievement is probably the greatest of which our age has to boast; and I know of no age (except perhaps the golden age of Greece) which has a more convincing proof to offer of the transcendent genius of its great men. Of the three problems, that of the infinitesimal was solved by Weierstrass; the solution of the other two was begun by Dedekind, and definitively accomplished by Cantor.

The infinitesimal played formerly a great part in mathematics. It was introduced by the Greeks, who regarded a circle as differing infinitesimally from a polygon with a very large number of very small equal sides. It gradually grew in importance, until, when Leibniz invented the Infinitesimal Calculus, it seems to become the fundamental notion of all higher mathematics. Carlyle tells, in his *Frederick the Great*, how Leibniz used to discourse to Queen Sophia Charlotte of Prussia concerning the infinitely little, and how she would reply that on that subject she needed no instruction – the behaviour of courtiers had made her thoroughly familiar with it. But philosophers and mathematicians – who for the most part had less acquaintance with courts – continued to discuss this topic, though without making any advance. The Calculus required continuity, and continuity was supposed to require the infinitely little; but nobody could discover what the infinitely little might be. It was plainly not quote zero, because a sufficiently large number of infinitesimals, added together, were seen to make up a finite whole. But nobody

---

[1] Professor of Mathematics in the University of Berlin. He died in 1897.

could point out any fraction which was not zero, and yet not finite. Thus there was a deadlock. But at last Weierstrass discovered that the infinitesimal was not needed at all, and that everything could be accomplished without it. Thus there was no longer any need to suppose that there was such a thing. Nowadays, therefore, mathematicians are more dignified than Leibniz: instead of talking about the infinitely small, they talk about the infinitely great – a subject which, however appropriate to monarchs, seems, unfortunately, to interest them even less than the infinitely little interested the monarchs to whom Leibniz discoursed.

The banishment of the infinitesimal has all sorts of odd consequences, to which one has to become gradually accustomed. For example, there is no such thing as the next moment. The interval between one moment and the next would have to be infinitesimal, since, if we take two moments with a finite interval between them, there are always other moments in the interval. Thus if there are to be no infinitesimals, no two moments are quite consecutive, but there are always other moments between any two. Hence, there must be an infinite number of moments between any two: because if there were a finite number one would be nearest the first of the two moments, and therefore next to it. This might be thought to be a difficulty; but, as a matter of fact, it is here that the philosophy of the infinite comes in, and makes all straight.

The same sort of thing happens in space. If any piece of matter be cut in two, and then each part be halved, and so on, the bits will become smaller and smaller, and can theoretically be made as small as we please. However small they may be, they can still be cut up and made smaller still. But they will always have *some* finite size, however small they may be. We never reach the infinitesimal in this way, and no finite number of divisions will bring us to points. Nevertheless there *are* points, only these are not to be reached by successive divisions. Here again, the philosophy of the infinite shows us how this is possible, and why points are not infinitesimal lengths.

As regards motion and change, we get similarly curious results. People used to think that when a thing changes, it must be in a state of change, and that when a thing moves, it is in a state of motion. This is now known to be a mistake. When a body moves, all that can be said is that it is in one place at one time and in another at another. We must not say that it will be in a neighbouring place at the next instant, since there is no next instant. Philosophers often tell us that when a body is in motion, it changes its position within the instant. To this view Zeno long ago made the fatal retort that every body always is where it is; but a retort so simple and brief was not the kind to which philosophers are accustomed to give weight, and they have continued down to our own day to repeat the same phrases which roused the Eleatic's destructive ardour. It was only recently that it became possible to explain motion in detail in accordance with Zeno's platitude, and in opposition to the philosopher's paradox. We may now at last indulge the comfortable belief that a body in motion is just as truly where it is as a body at rest. Motion consists merely in the fact that bodies are sometimes in one place and sometimes in another, and that they are at intermediate places at intermediate times. Only those who have waded through the quagmire of philosophic speculation on this subject can realise what a liberation from antique prejudices is involved in this simple and straightforward commonplace.

The philosophy of the infinitesimal, as we have just seen, is mainly negative. People used to believe in it, and now they have found out their mistake. The philosophy of the infinite, on the other hand, is wholly positive. It was formerly supposed that infinite numbers, and the mathematical infinite generally, were self-contradictory. But as it was obvious that there were infinities – for example, the number of numbers – the contradictions of infinity seemed unavoidable, and philosophy seemed to have wandered into a "cul-de-sac." This difficulty led to Kant's antinomies, and hence, more or less indirectly, to much of Hegel's dialectic method. Almost all current philosophy is upset by the fact (of which very few philosophers are as yet aware) that all the ancient and respectable contradictions in the notion of the infinite have been once for all disposed of. The method by which this has been done is most interesting and instructive. In the first place, though people had talked glibly about infinity every since the beginnings of Greek thought, nobody had ever thought of asking, What is infinity? If any philosopher had been asked for a definition of infinity, he might have produced some unintelligible rigmarole, but he would certainly not have been able to give a definition that had any meaning at all. Twenty years ago, roughly speaking, Dedekind and Cantor asked this question, and, what is more remarkable, they answered it. They found, that is to say, a perfectly precise definition of an infinite number or an infinite collection of things. This was the first and perhaps greatest step. It then remained to examine the supposed contradictions in this notion. Here Cantor proceeded in the only proper way. He took pairs of contradictory propositions, in which both sides of the contradiction would be usually regarded as demonstrable, and he strictly examined the supposed proofs. He found that all proofs adverse to infinity involved a certain principle, at first sight obviously true, but destructive, in its consequences, of almost all mathematics. The proofs favourable to infinity, on the other hand, involved no principle that had evil consequences. It thus appeared that common sense had allowed itself to be taken in by a specious maxim, and that, when once this maxim was rejected, all went well.

The maxim in question is, that if one collection is part of another, the one which is a part has fewer terms than the one of which it is a part. This maxim is true of finite numbers. For example, Englishmen are only some among Europeans, and there are fewer Englishmen than Europeans. But when we come to infinite numbers, this is no longer true. This breakdown of the maxim gives us the precise definition of infinity. A collection of terms is infinite when it contains as parts other collections which have just as many terms as it has. If you can take away some of the terms of a collection, without diminishing the number of terms, then there are an infinite number of terms in the collection. For example, there are just as many even numbers as there are numbers altogether, since every number can be doubled.

This may be seen by putting odd and even numbers together in one row, and even numbers alone in a row below: –

1, 2, 3, 4, 5, *ad infinitum*
2, 4, 6, 8, 10, *ad infinitum*

There are obviously just as many numbers in the row below as in the row above, because there is one below for each one above. This property, which was formerly thought to be a contradiction, is now transformed into a harmless definition of infinity, and shows, in the above case, that the number of finite numbers is infinite.

But the uninitiated may wonder how it is possible to deal with a number which cannot be counted. It is impossible to count up *all* the numbers, one by one, because, however many we may count, there are always more to follow. The fact is that counting is a very vulgar and elementary way of finding out how many terms there are in a collection. And in any case, counting gives us what mathematicians call the *ordinal* number of our terms; that is to say, it arranges our terms in an order or series, and its result tells us what type of series results from this arrangement. In other words, it is impossible to count things without counting some first and others afterwards, so that counting always has to do with order. Now when there are only a finite number of terms, we can count them in any order we like; but when there are an infinite number, what corresponds to counting will give us quite different results according to the way in which we carry out the operation. Thus the ordinal number, which results from what, in a general sense may be called counting, depends not only upon how many terms we have, but also (where the number of terms is infinite) upon the way in which the terms are arranged.

The fundamental infinite numbers are not ordinal, but are what is called *cardinal*. They are not obtained by putting our terms in order and counting them, but by a different method, which tells us, to begin with, whether two collections have the same number of terms, or, if not, which is the greater.[1] It does not tell us, in the way in which counting does, *what* number of terms a collection has; but if we define a number as the number of terms in such and such a collection, then this method enables us to discover whether some other collection that may be mentioned has more or fewer terms. An illustration will show how this is done. If there existed some country in which, for one reason or another, it was impossible to take a census, but in which it was known that every man had a wife and every woman a husband, then (provided polygamy was not a national institution) we should know, without counting, that there were exactly as many men as there were women in that country, neither more nor less. This method can be applied generally. If there is some relation which, like marriage, connects the things in one collection

---

1 [Note added in 1917.] Although some infinite numbers are greater than some others, it cannot be proved that of any two infinite numbers one must be the greater.

each with one of the things in another collection, and vice versa, then the two collections have the same number of terms. This was the way in which we found that there are as many even numbers as there are numbers. Every number can be doubled, and every even number can be halved, and each process gives just one number corresponding to the one that is doubled or halved. And in this way we can find any number of collections each of which has just as many terms as there are finite numbers. If every term of a collection can be hooked on to a number, and all the finite numbers are used once, and only once, in the process, then our collection must have just as many terms as there are finite numbers. This is the general method by which the numbers of infinite collections are defined.

But it must not be supposed that all infinite numbers are equal. On the contrary, there are infinitely more infinite numbers than finite ones. There are more ways of arranging the finite numbers in different types of series than there are finite numbers. There are probably more points in space and more moments in time than there are finite numbers. There are exactly as many fractions as whole numbers, although there are an infinite number of fractions between any two whole numbers. But there are more irrational numbers than there are whole numbers or fractions. There are probably exactly as many points in space as there are irrational numbers, and exactly as many points on a line a millionth of an inch long as in the whole of infinite space. There is a greatest of all infinite numbers, which is the number of things altogether, of every sort and kind. It is obvious that there cannot be a greater number than this, because, if everything has been taken, there is nothing left to add. Cantor has a proof that there is no greatest number, and if this proof were valid, the contradictions of infinity would reappear in a sublimated form. But in this one point, the master has been guilty of a very subtle fallacy, which I hope to explain in some future work.[1]

We can now understand why Zeno believed that Achilles cannot overtake the tortoise and why as a matter of fact he can overtake it. We shall see that all the people who disagreed with Zeno had no right to do so, because they all accepted premises from which his conclusion followed. The argument is this: Let Achilles and the tortoise start along a road at the same time, the tortoise (as is only fair) being allowed a handicap. Let Achilles go twice as fast as the tortoise, or ten times or a hundred times as fast. Then he will never reach the tortoise. For at every moment the tortoise is somewhere and Achilles is somewhere; and neither is ever twice in the same place while the race is going on. Thus the tortoise goes to just as many places as Achilles does, because each is in one place at one moment, and in another at any other moment. But if Achilles were to catch up with the tortoise, the places where the tortoise would have been would be only part of the places

---

[1] Cantor was not guilty of a fallacy on this point. His proof that there is no greatest number is still valid. The solution of the puzzle is complicated and depends upon the theory of types, which is explained in *Principia Mathematica*, Vol. 1 (Camb. Univ. Press, 1910). [Note added in 1917.]

where Achilles would have been. Here, we must suppose, Zeno appealed to the maxim that the whole has more terms than the part.[1] Thus if Achilles were to overtake the tortoise, he would have been in more places than the tortoise; but we saw that he must, in any period, be in exactly as many places as the tortoise. Hence we infer that he can never catch the tortoise. This argument is strictly correct, if we allow the axiom that the whole has more terms than the part. As the conclusion is absurd, the axiom must be rejected, and then all goes well. But there is no good word to be said for the philosophers of the past two thousand years and more, who have all allowed the axiom and denied the conclusion.

The retention of this axiom leads to absolute contradictions, while its rejection leads only to oddities. Some of these oddities, it must be confessed, are very odd. One of them, which I call the paradox of Tristram Shandy, is the converse of the Achilles, and shows that the tortoise, if you give him time, will go just as far as Achilles. Tristram Shandy, as we know, employed two years in chronicling the first two days of his life, and lamented that, at this rate, material would accumulate faster than he could deal with it, so that, as years went by, he would be farther and farther from the end of his history. Now I maintain that, if he had lived for ever, and had not wearied of his task, then, even if his life had continued as it began, no part of his biography would have remained unwritten. For consider: the hundredth day will be described in the hundredth year, the thousandth in the thousandth year, and so on. Whatever day we may choose as so far on that he cannot hope to reach it, that day will be described in the corresponding year. Thus any day that may be mentioned will be written up sooner or later, and therefore no part of the biography will remain permanently unwritten. This paradoxical but perfectly true proposition depends upon the fact that the number of days in all time is no greater than the number of years.

Thus on the subject of infinity it is impossible to avoid conclusions which at first sight appear paradoxical, and this is the reason why so many philosophers have supposed that there were inherent contradictions in the infinite. But a little practice enables one to grasp the true principles of Cantor's doctrine, and to acquire new and better instincts as to the true and the false. The oddities then become no odder than the people at the antipodes, who used to be thought impossible because they would find it so inconvenient to stand on their heads.

The solution of the problems concerning infinity has enabled Cantor to solve also the problems of continuity. Of this, as of infinity, he has given a perfectly precise definition, and has shown that there are no contradictions in the notion so

---

1 This must not be regarded as a historically correct account of what Zeno actually had in mind. It is a new argument for his conclusion, not the argument which influenced him. On this point, see e.g., C.D. Broad, "Note on Achilles and the Tortoise," *Mind*, N.S., Vol. XXII, pp. 318–19. Much valuable work on the interpretation of Zeno has been done since this article was written. [Note added in 1917.]

defined. But this subject is so technical that it is impossible to give any account of it here.

The notion of continuity depends upon that of *order*, since continuity is merely a particular type of order. Mathematics has, in modern times, brought order into greater and greater prominence. In former days, it was supposed (and philosophers are still apt to suppose) that quantity was the fundamental notion of mathematics. But nowadays, quantity is banished altogether, except from one little corner of Geometry, while order more and more reigns supreme. The investigation of different kinds of series and their relations is now a very large part of mathematics, and it has been found that this investigation can be conducted without any reference to quantity, and, for the most part, without any reference to number. All types of series are capable of formal definition, and their properties can be deduced from the principles of symbolic logic by means of the Algebra of Relatives. The notion of a limit, which is fundamental in the greater part of higher mathematics, used to be defined by means of quantity, as a term to which the terms of some series approximate as nearly as we please. But nowadays the limit is defined quite differently, and the series which it limits may not approximate to it at all. This improvement also is due to Cantor, and it is one which has revolutionized mathematics. Only order is now relevant to limits. Thus, for instance, the smallest of the infinite integers is the limit of the finite integers, though all finite integers are at an infinite distance from it. The study of different types of series is a general subject of which the study of ordinal numbers (mentioned above) is a special and very interesting branch. But the unavoidable technicalities of this subject render it impossible to explain to any but professed mathematicians.

Geometry, like Arithmetic, has been subsumed, in recent times, under the general study of order. It was formerly supposed that Geometry was the study of the nature of the space in which we live, and accordingly it was urged, by those who held that what exists can only be known empirically, that Geometry should really be regarded as belonging to applied mathematics. But it has gradually appeared, by the increase of non-Euclidean systems, that Geometry throws no more light upon the nature of space than Arithmetic throws upon the population of the United States. Geometry is a whole collection of deductive sciences based on a corresponding collection of sets of axioms. One set of axioms is Euclid's; other equally good sets of axioms lead to other results. Whether Euclid's axioms are true, is a question as to which the pure mathematician is indifferent; and, what is more, it is a question which it is theoretically impossible to answer with certainty in the affirmative. It might possibly be shown, by very careful measurements, that Euclid's axioms are false; but no measurements could ever assure us (owing to the errors of observation) that they are exactly true. Thus the geometer leaves to the man of science to decide, as best he may, what axioms are most nearly true in the actual world. The geometer takes any set of axioms that seem interesting, and deduces their consequences. What defines Geometry, in this sense, is that the axioms must give rise to a series of more than one dimension. And it is thus that Geometry becomes a department in the study of order.

In Geometry, as in other parts of mathematics, Peano and his disciples have done work of the very greatest merit as regards principles. Formerly, it was held by philosophers and mathematicians alike that the proofs in Geometry depended on the figure; nowadays, this is known to be false. In the best books there are no figures at all. The reasoning proceeds by the strict rules of formal logic from a set of axioms laid down to begin with. If a figure is used, all sorts of things seem obviously to follow, which no formal reasoning can prove from the explicit axioms, and which, as a matter of fact, are only accepted because they are obvious. By banishing the figure, it becomes possible to discover *all* the axioms that are needed; and in this way all sorts of possibilities, which would have otherwise remained undetected, are brought to light.

One great advance, from the point of view of correctness, has been made by introducing points as they are required, and not starting, as was formerly done, by assuming the whole of space. This method is due partly to Peano, partly to another Italian named Fano. To those unaccustomed to it, it has an air of somewhat wilful pedantry. In this way, we begin with the following axioms: (1) There is a class of entities called *points*. (2) There is at least one point. (3) If a be a point, there is at least one other point besides a. Then we bring in the straight line joining two points, and begin again with (4) namely, on the straight line joining a and b, there is at least one other point besides a and b. (5) There is at least one point not on the line ab. And so we go on, till we have the means of obtaining as many points as we require. But the word *space*, as Peano humorously remarks, is one for which Geometry has no use at all.

The rigid methods employed by modern geometers have deposed Euclid from his pinnacle of correctness. It was thought, until recent times, that, as Sir Henry Savile remarked in 1621, there were only two blemishes in Euclid, the theory of parallels and the theory of proportion. It is now known that these are almost the only points in which Euclid is free from blemish. Countless errors are involved in his first eight propositions. That is to say, not only is it doubtful whether his axioms are true, which is a comparatively trivial matter, but it is certain that his propositions do not follow from the axioms which he enunciates. A vastly greater number of axioms, which Euclid unconsciously employs, are required for the proof of his propositions. Even in the first proposition of all, where he constructs an equilateral triangle on a given base, he uses two circles which are assumed to intersect. But no explicit axiom assures us that they do so, and in some kinds of spaces they do not always intersect. It is quite doubtful whether our space belongs to one of these kinds or not. Thus Euclid fails entirely to prove his point in the very first proposition. As he is certainly not an easy author, and is terribly long-winded, he has no longer any but an historical interest. Under these circumstances, it is nothing less than a scandal that he should still be taught to boys in

England.[1] A book should have either intelligibility or correctness; to combine the two is impossible, but to lack both is to be unworthy of such a place as Euclid has occupied in education.

The most remarkable result of modern methods in mathematics is the importance of symbolic logic and of rigid formalism. Mathematicians, under the influence of Weierstrass, have shown in modern times a care for accuracy, and an aversion to slipshod reasoning, such as had not been known among them previously since the time of the Greeks. The great inventions of the seventeenth century – Analytical Geometry and the Infinitesimal Calculus – were so fruitful in new results that mathematicians had neither time nor inclination to examine their foundations. Philosophers, who should have taken up the task, had too little mathematics which have now been found necessary for any adequate discussion. Thus mathematicians were only awakened from their "dogmatic slumbers" when Weierstrass and his followers showed that many of their most cherished propositions are in general false. Macaulay, contrasting the certainty of mathematics with the uncertainty of philosophy, asks who ever heard of a reaction against Taylor's theorem? If he had lived now, he himself might have heard of such a reaction, for this is precisely one of the theorems which modern investigations have overthrown. Such rude shocks to mathematical faith have produced that love of formalism which appears, to those who are ignorant of its motive, to be mere outrageous pedantry.

The proof that all pure mathematics, including Geometry, is nothing but formal logic, is a fatal blow to the Kantian philosophy. Kant, rightly perceiving that Euclid's propositions could not be deduced from Euclid's axioms without the help of the figures, invented a theory of knowledge to account for this fact; and it accounted so successfully that, when the fact is shown to be a mere defect in Euclid, and not a result of the nature of geometrical reasoning, Kant's theory also has to be abandoned. The whole doctrine of *a priori* intuitions, by which Kant explained the possibility of pure mathematics, is wholly inapplicable to mathematics in its present form. The Aristotelian doctrines of the schoolmen come nearer in spirit to the doctrines which modern mathematics inspire; but the schoolmen were hampered by the fact that their formal logic was very defective, and that the philosophical logic based upon the syllogism showed a corresponding narrowness. What is now required is to give the greatest possible development to mathematical logic, to allow to the full the importance of relations, and then to found upon this secure basis a new philosophical logic, which may hope to borrow some of the exactitude and certainty of its mathematical foundation. If this can be successfully accomplished, there is every reason to hope that the near future will be as great an epoch in pure philosophy as the immediate past has been

---

1 Since the above was written, he has ceased to be used as a textbook. But I fear many of the books now used are so bad that the change is no great improvement. [Note added in 1917.]

in the principles of mathematics. Great triumphs inspire great hopes; and pure thought may achieve, within our generation, such results as will place our time, in this respect, on a level with the greatest age of Greece.[1]

---

[1] The greatest age of Greece was brought to an end by the Peloponnesian War. [Note added in 1917.]

11
ON DENOTING

By a 'denoting phrase' I mean a phrase such as any one of the following: a man, some man, any man, every man, all men, the present King of England, the present King of France, the centre of mass of the solar system at the first instant of the twentieth century, the revolution of the earth round the sun, the revolution of the sun round the earth. Thus a phrase is denoting solely in virtue of its *form*. We may distinguish three cases: (1) A phrase may be denoting, and yet not denote anything; e.g., 'the present King of France'. (2) A phrase may denote one definite object; e.g., 'the present King of England' denotes a certain man. (3) A phrase may denote ambiguously; e.g., 'a man' denotes not many men, but an ambiguous man. The interpretation of such phrases is a matter of considerable difficulty; indeed, it is very hard to frame any theory not susceptible of formal refutation. All the difficulties with which I am acquainted are met, so far as I can discover, by the theory which I am about to explain.

The subject of denoting is of very great importance, not only in logic and mathematics, but also in theory of knowledge. For example, we know that the centre of mass of the solar system at a definite instant is some definite point, and we can affirm a number of propositions about it; but we have no immediate *acquaintance* with this point, which is only known to us by description. The distinction between *acquaintance* and *knowledge about* is the distinction between the things we have presentations of, and the things we only reach by means of denoting phrases. It often happens that we know that a certain phrase denotes unambiguously, although we have no acquaintance with what it denotes; this occurs in the above case of the centre of mass. In perception we have acquaintance with the objects of perception, and in thought we have acquaintance with objects of a more abstract logical character; but we do not necessarily have acquaintance with the objects denoted by phrases composed of words with whose meanings we are acquainted. To take a very important instance: there seems no reason to believe that we are ever acquainted with other people's minds, seeing that these are not directly perceived; hence what we know about them is obtained through denoting. All thinking has to start from acquaintance; but it succeeds in thinking *about* many things with which we have no acquaintance.

The course of my argument will be as follows. I shall begin by stating the theory I intend to advocate;* I shall then discuss the theories of Frege and Meinong, showing why neither of them satisfies me; then I shall give the grounds in favour of my theory; and finally I shall briefly indicate the philosophical consequences of my theory.

My theory, briefly, is as follows. I take the notion of the *variable* as fundamental; I use '$C(x)$' to mean a proposition† in which x is a constituent, where x, the variable, is essentially and wholly undetermined. Then we can consider the two notions '$C(x)$ is always true' and '$C(x)$ is sometimes true'‡. Then *everything* and *nothing* and *something* (which are the most primitive of denoting phrases) are to be interpreted as follows:

C (everything) means '$C(x)$ is always true';
C (nothing) means ' "$C(x)$ is false" is always true';
C (something) means 'It is false that "$C(x)$ is false" is always true'.§

Here the notion '$C(x)$ is always true' is taken as ultimate and indefinable, and the others are defined by means of it. *Everything, nothing,* and *something* are not assumed to have any meaning in isolation, but a meaning is assigned to *every* proposition in which they occur. This is the principle of the theory of denoting I wish to advocate: that denoting phrases never have any meaning in themselves, but that every proposition in whose verbal expression they occur has a meaning. The difficulties concerning denoting are, I believe, all the result of a wrong analysis of propositions whose verbal expressions contain denoting phrases. The proper analysis, if I am not mistaken, may be further set forth as follows.

Suppose now we wish to interpret the proposition, 'I met a man'. If this is true, I met some definite man; but that is not what I affirm. What I affirm is, according to the theory I advocate:

'"I met x, and x is human" is not always false'.

Generally, defining the class of men as the class of objects having the predicate *human*, we say that:

---

* I have discussed this subject in Principles of Mathematics, Chap. V, and § 476. The theory there advocated in very nearly the same as Frege's, and is quite different from the theory to be advocated in what follows.
† More exactly, a propositional function.
‡ The second of these can be defined by means of the first, if we take it to mean, 'It is not true that "$C(x)$ is false" is always true'.
§ I shall sometimes use, instead of this complicated phrase, the phrase '$C(x)$ is not always false', or '$C(x)$ is sometimes true', supposed defined to mean the same as the complicated phrase.

'C (a man)' means ' "$C(x)$ and x is human" is not always false'.

This leaves 'a man', by itself, wholly destitute of meaning, but gives a meaning to every proposition in whose verbal expression 'a man' occurs.

Consider next the proposition 'all men are mortal'. This proposition* is really hypothetical and states that *if* anything is a man, it is mortal. That is, it states that if x is a man, x is mortal, whatever x may be. Hence, substituting 'x is human' for 'x is a man', we find:

'All men are mortal' means ' "If x is human, x is mortal" is always true'.

This is what is expressed in symbolic logic by saying that 'all men are mortal' means ' "x is human" implies "x is mortal" for all values of x'. More generally, we say:

'C (all men)' means ' "If x is human, then $C(x)$ is true" is always true'.

Similarly

'C (no men)' means ' "If x is human, the $C(x)$ is false" is always true'.
'C (some men)' will mean the same as 'C (a man)',† and
'C (a man)' means 'It is false that "$C(x)$ and x is human" is always false'.
'C (every man)' will mean the same as 'C (all men)'.

It remains to interpret phrases containing *the*. These are by far the most interesting and difficult of denoting phrases. Take as an instance 'the father of Charles II was executed'. This asserts that there was an x who was the father of Charles II and was executed. Now *the*, when it is strictly used, involves uniqueness; we do, it is true, speak of '*the* son of So-and-so' even when So-and-so has several sons, but it would be more correct to say '*a* son of So-and-so'. Thus for our purposes we take *the* as involving uniqueness. Thus when we say 'x was *the* father of Charles II' we not only assert that x had a certain relation to Charles II, but also that nothing else had this relation. The relation in question, without the assumption of uniqueness, and without any denoting phrases, is expressed by 'x begat Charles II'. To get an equivalent of 'x was the father of Charles II', we must add, 'If y is other than x, y did not beget Charles II', or, what is equivalent, 'If y begat Charles II, y is identical with x'. Hence 'x is the father of Charles II' becomes: 'x begat Charles II; and "if y begat Charles II, y is identical with x" is always true of y'.

---

* As has been ably argued in Mr. Bradley's *Logic*, Book I, Chap. II.
† Psychologically 'C (a man)' has a suggestion of *only one*, and 'C (some men)' has a suggestion of *more than one*; but we may neglect these suggestions in a preliminary sketch.

Thus, 'the father of Charles II was executed' becomes:

'It is not always false of x that x begat Charles II and that x was executed and that "if y begat Charles II, y is identical with x" is always true of y'.

This may seem a somewhat incredible interpretation; but I am not at present giving reasons, I am merely *stating* the theory.

To interpret 'C (the father of Charles II)', where C stands for any statement about him, we have only to substitute $C(x)$ for 'x was executed' in the above. Observe that, according to the above interpretation, whatever statement C may be, 'C (the father of Charles II)' implies:

'It is not always false of x that "if y begat Charles II, y is identical with x" is always true of y',

which is what is expressed in common language by 'Charles II had one father and no more'. Consequently if this condition fails, *every* proposition of the form 'C (the father of Charles II)' is false. Thus e.g. every proposition of the form 'C (the present King of France)' is false. This is a great advantage in the present theory. I shall show later that it is not contrary to the law of contradiction, as might be at first supposed.

The above gives a reduction of all propositions in which denoting phrases occur to forms in which no such phrases occur. Why it is imperative to effect such a reduction, the subsequent discussion will endeavour to show.

The evidence for the above theory is derived from the difficulties which seem unavoidable if we regard denoting phrases as standing for genuine constituents of the propositions in whose verbal expressions they occur. Of the possible theories which admit such constituents the simplest is that of Meinong.* This theory regards any grammatically correct denoting phrase as standing for an *object*. Thus 'the present King of France', 'the round square', etc., are supposed to be genuine objects. It is admitted that such objects do not *subsist*, but nevertheless they are supposed to be objects. This is in itself a difficult view; but the chief objection is that such objects, admittedly, are apt to infringe the law of contradiction. It is contended, for example, that the existent present King of France exists, and also does not exist; that the round square is round, and also not round, etc. But this is intolerable; and if any theory can be found to avoid this result, it is surely to be preferred.

The above breach of the law of contradiction is avoided by Frege's theory. He distinguishes, in a denoting phrase, two elements, which we may call the *meaning* and the *denotation*.† Thus 'the centre of mass of the solar system at the beginning

---

* See *Untersuchungen zur Gegenstandstheorie une Psychologie* (Leipzig, 1904) the first three articles (by Meinong, Ameseder and Mally respectively).
† See his 'Sense and Reference' 100.

of the twentieth century' is highly complex in *meaning*, but its *denotation* is a certain point, which is simple. The solar system, the twentieth century, etc., are constituents of the *meaning*; but the *denotation* has no constituents at all.* One advantage of this distinction is that it shows why it is often worth while to assert identity. If we say 'Scott is the author of *Waverley*', we assert an identity of denotation with a difference of meaning. I shall, however, not repeat the grounds in favour of this theory, as I have urged its claims elsewhere (loc. cit.), and am now concerned to dispute those claims.

One of the first difficulties that confront us, when we adopt the view that denoting phrases *express* a meaning and *denote* a denotation,† concerns the cases in which the denotation appears to be absent. If we say 'the King of England is bald', that is, it would seem, not a statement about the complex *meaning* 'the King of England', but about the actual man denoted by the meaning. But now consider 'the King of France is bald'. By parity of form, this also ought to be about the denotation of the phrase 'the King of France'. But this phrase, though it has a *meaning* provided 'the King of England' has a meaning, has certainly no denotation, at least in any obvious sense. Hence one would suppose that 'the King of France is bald' ought to be nonsense; but it is not nonsense, since it is plainly false. Or again consider such a proposition as the following: 'If u is a class which has only one member, then that one member is a member of u', or, as we may state it, 'If u is a unit class, the u is a u'. This proposition ought to be *always* true, since the conclusion is true whenever the hypothesis is true. But 'the u' is a denoting phrase, and it is the denotation, not the meaning, that is said to be a u. Now if u is *not* a unit class, 'the u' seems to denote nothing; hence our proposition would seem to become nonsense as soon as u is not a unit class.

Now it is plain that such propositions do *not* become nonsense merely because their hypotheses are false. The King in *The Tempest* might say, 'If Ferdinand is not drowned, Ferdinand is my only son'. Now 'my only son' is a denoting phrase, which, on the face of it, has a denotation when, and only when, I have exactly one son. But the above statement would nevertheless have remained true if Ferdinand had been in fact drowned. Thus we must either provide a denotation in cases in which it is at first sight absent, or we must abandon the view that the denotation is what is concerned in propositions which contain denoting phrases. The latter is the course that I advocate. The former course may be taken, as by Meinong, by admit-

---

* Frege distinguishes the two elements of meaning and denotation everywhere, and not only in complex denoting phrases. Thus it is the *meanings* of the constituents of a denoting complex that enter into its *meaning*, not their *denotation*. In the proposition 'Mont Blanc is over 1,000 metres high', it is, according to him, the *meaning* of 'Mont Blanc', not the actual mountain, that is a constituent of the *meaning* of the proposition.
† In this theory, we shall say that the denoting phrase *expresses* a meaning; and we shall say both of the phrase and of the meaning that they *denote* a denotation. In the other theory, which I advocate, there is no *meaning*, and only sometimes a *denotation*.

ting objects which do not subsist, and denying that they obey the law of contradiction; this, however, is to be avoided if possible. Another way of taking the same course (so far as our present alternative is concerned) is adopted by Frege, who provides by definition some purely conventional denotation for the cases in which otherwise there would be none. Thus 'the King of France', is to denote the null-class; 'the only son of Mr. So-and-so' (who has a fine family of ten), is to denote the class of all his sons; and so on. But this procedure, though it may not lead to actual logical error, is plainly artificial and does not give an exact analysis of the matter. Thus is we allow that denoting phrases, in general, have the two sides of meaning and denotation, the cases where there seems to be no denotation cause difficulties both on the assumption that there really is a denotation and on the assumption that there really is none.

A logical theory may be tested by its capacity for dealing with puzzles, and it is a wholesome plan, in thinking about logic, to stock the mind with as many puzzles as possible, since these serve much the same purpose as is served by experiments in physical science. I shall therefore state three puzzles which a theory as to denoting ought to be able to solve; and I shall show later that my theory solves them.

(1) If a is identical with b, whatever is true of the one is true of the other, and either may be substituted for the other in any proposition without altering the truth or falsehood of that proposition. Now George IV wished to know whether Scott was the author of *Waverley*; and in fact Scott *was* the author of *Waverley*. Hence we may substitute *Scott* for *the author of 'Waverley'*, and thereby prove that George IV wished to know whether Scott was Scott. Yet an interest in the law of identity can hardly be attributed to the first gentleman of Europe.

(2) By the law of excluded middle, either 'A is B' or 'A is not B' must be true. Hence either 'the present King of France is bald' or 'the present King of France is not bald' must be true. Yet if we enumerated the things that are bald, and then the things that are not bald, we should not find the present King of France in either list. Hegelians, who love a synthesis, will probably conclude that he wears a wig.

(3) Consider the proposition 'A differs from B'. If this is true, there is a difference between A and B, which fact may be expressed in the form 'the difference between A and B subsists'. But if it is false that A differs from B, then there is no difference between A and B, which fact may be expressed in the form 'the difference between A and B does not subsist'. But how can a non-entity be the subject of a proposition? 'I think, therefore I am' is no more evident than 'I am the subject of a proposition, therefore I am', provided 'I am' is taken to assert subsistence or being,* not existence. Hence, it would appear, it must always be self-contradictory to deny the being of anything; but we have seen, in connexion with Meinong, that to admit being also sometimes leads to contradictions. Thus if A and B do not

---

* I use these as synonyms.

differ, to suppose either that there is, or that there is not, such an object as 'the difference between *A* and *B*' seems equally impossible.

The relation of the meaning to the denotation involves certain rather curious difficulties, which seem in themselves sufficient to prove that the theory which leads to such difficulties must be wrong.

When we wish to speak about the *meaning* of a denoting phrase, as opposed to its *denotation*, the natural mode of doing so is by inverted commas. Thus we say:

The centre of mass of the solar system is a point, not a denoting complex;
'The centre of mass of the solar system' is a denoting complex, not a point.

Or again,
The first line of Gray's Elegy states a proposition.

'The first line of Gray's Elegy' does not state a proposition. Thus taking any denoting phrase, say *C* and '*C*', where the difference of the two is of the kind exemplified in the above two instances.

We say, to begin with, that when *C* occurs it is the *denotation* that we are speaking about; but when '*C*' occurs, it is the *meaning*. Now the relation of meaning and denotation is not merely linguistic through the phrase: there must be a logical relation involved, which we express by saying that the meaning denotes the denotation. But the difficulty which confronts us is that we cannot succeed in *both* preserving the connexion of meaning and denotation *and* preventing them from being one and the same; also that the meaning cannot be got at except by means of denoting phrases. This happens as follows.

The one phrase *C* was to have both meaning and denotation. But if we speak of 'the meaning of *C*', that gives us the meaning (if any) of the denotation. 'The meaning of the first line of Gray's Elegy' is the same as 'The meaning of "The curfew tolls the knell of parting day",' and is not the same as 'The meaning of "the first line of Gray's Elegy".' Thus in order to get the meaning we want, we must speak not of 'the meaning of *C*', but of 'the meaning of "*C*",' which is the same as '*C*' by itself. Similarly 'the denotation of *C*' does not mean the denotation we want, but means something which, if it denotes at all, denotes what is denoted by the denotation we want. For example, let '*C*' be 'the denoting complex occurring the second of the above instances'. Then

C = 'the first line of Gray's Elegy', and

the denotation of *C* = The curfew tolls the knell of parting day. But what we *meant* to have as the denotation was 'the first line of Gray's Elegy'. Thus we have failed to get what we wanted.

The difficulty in speaking of the meaning of a denoting complex may be stated thus: The moment we put the complex in a proposition, the proposition is about the denotation; and if we make a proposition in which the subject is 'the meaning of *C*', then the subject is the meaning (if any) of the denotation, which was not intended. This leads us to say that, when we distinguish meaning and denotation, we must be dealing with the meaning: the meaning has denotation and is a complex, and there is not something Other than the meaning, which can be called the complex, and be said to *have* both meaning and denotation. The right phrase, on the view in question, is that some meanings have denotations.

But this only makes our difficulty in speaking of meanings more evident. For suppose *C* is our complex; then we are to say that *C is* the meaning of the complex. Nevertheless, whenever *C* occurs without inverted commas, what is said is not true of the meaning, but only of the denotation, as when we say: The centre of mass of the solar system is a point. Thus to speak of *C* itself, i.e., to make a proposition about the meaning, our subject must not be *C*, but something which denotes *C*. Thus '*C*', which is what we use when we want to speak of the meaning, must be not the meaning, but something which denotes the meaning. And *C* must not be a constituent of this complex (as it is of 'the meaning of *C*'); for if *C* occurs in the complex, it will be its denotation, not its meaning, that will occur, and there is no backward road from denotations to meanings, because every object can be denoted by an infinite number of different denoting phrases.

Thus it would seem that '*C*' and *C* are different entities, such that '*C*' denotes *C* ; but this cannot be an explanation, because the relation of '*C*' to *C* remains wholly mysterious; and where are we to find the denoting complex '*C*' which is to denote *C*? Moreover, when *C* occurs in a proposition, it is not *only* the denotation that occurs (as we shall see in the next paragraph); yet, on the view in question, *C* is only the denotation, the meaning being wholly relegated to '*C*'. This is an inextricable tangle, and seems to prove that the whole distinction of meaning and denotation has been wrongly conceived.

That the meaning is relevant when a denoting phrase occurs in a proposition is formally proved by the puzzle about the author of *Waverley*. The proposition 'Scott was the author of *Waverley*' has a property not possessed by 'Scott was Scott', namely the property that George IV wished to know whether it was true. Thus the two are not identical propositions; hence the meaning of 'the author of *Waverley*' must be relevant as well as the denotation, if we adhere to the point of view to which this distinction belongs. Yet, as we have just seen, so long as we adhere to this point of view, we are compelled to hold that only the denotation can be relevant. Thus the point of view in question must be abandoned.

It remains to show how all the puzzles we have been considering are solved by the theory explained at the beginning of this article.

According to the view which I advocate, a denoting phrase is essentially *part* of a sentence, and does not, like most single words, have any significance on its own account. If I say 'Scott was a man', that is a statement of the form 'x was a man', and it has 'Scott' for its subject. But if I say 'the author of *Waverley* was a man', that is not a statement of the form 'x was a man', and does not have 'the author of *Waverley*' for its subject. Abbreviating the statement made at the beginning of this article, we may put, in place of 'the author of *Waverley* was a man', the following: 'One and only one entity wrote *Waverley*, and that one was a man'. (This is not so strictly what is meant as what was said earlier; but it is easier to follow.) And speaking generally, suppose we wish to say that the author of *Waverley* had the property ϕ, what we wish to say is equivalent to 'One and only one entity wrote *Waverley*, and that one had the property ϕ'.

The explanation of *denotation* is now as follows. Every proposition in which 'the author of *Waverley*' occurs being explained as above, the proposition 'Scott was the author of *Waverley*' (i.e. 'Scott was identical with the author of *Waverley*') becomes 'One and only one entity wrote *Waverley*, and Scott was identical with that one'; or, reverting to the wholly explicit form: 'It is not always false of x that x wrote *Waverley*, that it is always true of y that if y wrote *Waverley* y is identical with x, and that Scott is identical with x'. Thus if 'C' is a denoting phrase, it may happen that there is one entity x (there cannot be more than one) for which the proposition 'x is identical with C' is true, this proposition being interpreted as above. We may then say that the entity x is the denotation of the phrase 'C'. Thus Scott is the denotation of 'the author of *Waverley*'. The 'C' in inverted commas will be merely the *phrase*, not anything that can be called the *meaning*. The phrase *per se* has no meaning, because in any proposition in which it occurs the proposition, fully expressed, does not contain the phrase, which has been broken up.

The puzzle about George IV's curiosity is now seen to have a very simple solution. The proposition 'Scott was the author of *Waverley*', which was written out in its unabbreviated form in the preceding paragraph, does not contain any constituent 'the author of *Waverley*' for which we could substitute 'Scott'. This does not interfere with the truth of inferences resulting from making what is *verbally* the substitution of 'Scott' for 'the author of *Waverley*', so long as 'the author of *Waverley*' has what I call a *primary* occurrence in the proposition considered. The difference of primary and secondary occurrences of denoting phrases is as follows:

When we say: 'George IV wished to know whether so-and-so', or when we say 'So-and-so is surprising' or 'So-and-so is true', etc., the 'so-and-so' must be a proposition. Suppose now that 'so-and-so' contains a denoting phrase. We may either eliminate this denoting phrase from the subordinate proposition 'so-and-so', or from the whole proposition in which 'so-and-so' is a mere constituent. Different propositions result according to which we do. I have heard of a touchy owner of a yacht to whom a guest, on first seeing it, remarked, 'I thought your yacht was larger than it is'; and the owner replied 'No, my yacht is not larger than it is'. What the guest meant was, 'The size that I thought your yacht was is greater than the size your yacht is'; the meaning attributed to him is, 'I thought the size of your yacht was greater than the size of our yacht'. To return to George IV and *Waverley*, when we say, 'George IV wished to know whether Scott was the author of *Waverley*', we normally mean 'George IV wished to know whether one and only one man wrote *Waverley* and Scott was that man'; but we *may* also mean: 'One and only one man wrote *Waverley*, and George IV wished to know whether Scott was that man'. In the latter, 'the author of *Waverley*' has a *primary* occurrence; in the former, a *secondary*. The latter might be expressed by 'George IV wished to know, concerning the man who in fact wrote *Waverley*, whether he was Scott'. This would be true, for example, if George IV had seen Scott at a distance, and had asked 'Is that Scott?' A *secondary* occurrence of a denoting phrase may be defined as one in which the phrase occurs in a proposition p which is a mere constituent of the proposition we are considering, and the substitution for the denoting phrase is to be effected in p, not in the whole proposition concerned. The ambiguity as between primary and secondary occurrences is hard to avoid in language; but it does no harm if we are on our guard against it. In symbolic logic it is of course easily avoided.

The distinction of primary and secondary occurrences also enables us to deal with the question whether the present King of France is bald or not bald, and generally with the logical status of denoting phrases that denote nothing. If 'C' is a denoting phrase, say 'the term having the property F', then

'C has the property ϕ' means 'one and only one term has the property F, and that one has the property ϕ'.*

If now the property F belongs to no terms, or to several, it follows that 'C has the property ϕ' is false for *all* values of ϕ. Thus 'the present King of France is bald' is certainly false; and 'the present King of France is not bald' is false if it means

'There is an entity which is now King of France and is not bald',

but is true if it means

'It is false that there is an entity which is now King of France and is bald'.

That is, 'the King of France is not bald' is false if the occurrence of 'the King of France' is *primary*, and true if it is *secondary*. Thus all propositions in which 'the King of France' has a primary occurrence are false; the denials of such propositions are true, but in them 'the King of France' has a secondary occurrence. Thus we escape the conclusion that the King of France has a wig.

We can now see also how to deny that there is such an object as the difference between A and B in the case when A and B do not differ. If A and B do differ, there is one and only one entity x such that 'x is the difference between A and B' is a true proposition; if A and B do not differ, there is no such entity x. Thus according to the meaning of denotation lately explained, 'the difference between A and B' has

---

* This is the abbreviated, not the stricter, interpretation.

a denotation when *A* and *B* differ, but not otherwise. This difference applies to true and false propositions generally. If '*a R b*' stands for '*a* has the relation *R* to *b*', then when *a R b* is true, there is such an entity as the relation *R* between *a* and *b*; when *a R b* is false, there is no such entity. Thus out of any proposition we can make a denoting phrase, which denotes an entity if the proposition is true, but does not denote an entity if the proposition is false. E.g., it is true (at least we will suppose so) that the earth revolves round the sun, and false that the sun revolves round the earth; hence 'the revolution of the earth round the sun' denotes an entity, while 'the revolution of the sun round the earth' does not denote an entity.*

The whole realm of non-entities, such as 'the round square', 'the even prime other than 2', 'Apollo', 'Hamlet', etc., can now be satisfactorily dealt with. All these are denoting phrases which do not denote anything. A proposition about Apollo means what we get by substituting what the classical dictionary tells us is meant by Apollo, say 'the sun-god'. All propositions in which Apollo occurs are to be interpreted by the above rules for denoting phrases. If 'Apollo' has a primary occurrence, the proposition containing the occurrence is false; if the occurrence is secondary, the proposition may be true. So again 'the round square is round' means 'there is one and only one entity *x* which is round and square, and that entity is round', which is a false proposition, not, as Meinong maintains, a true one. 'The most perfect Being has all perfections; existence is a perfection; therefore the most perfect Being exists' becomes:

'There is one and only one entity *x* which is most perfect; that one has all perfections; existence is a perfection; therefore that one exists.'

As a proof, this fails for want of a proof of the premiss

'there is one and only one entity *x* which is most perfect'.†

Mr. McColl (*Mind*, N.S., No. 54, and again No. 55, page 401) regards individuals as of two sorts, real and unreal; hence he defines the null-class as the class consisting of all unreal individuals. This assumes that such phrases as 'the present King of France', which do not denote a real individual, do, nevertheless, denote an individual, but an unreal one. This is essentially Meinong's theory, which we have seen reason to reject because it conflicts with the law of contradiction. With our theory of denoting, we are able to hold that there are no unreal individuals; so that

---

* The propositions from which such entities are derived are not identical either with these entities or with the propositions that these entities have being.
† The argument can be made to prove validity that all members of the class of most perfect Beings exist; it can also be proved formally that this class cannot have more than one member; but, taking the definition of perfection as possession of all positive predicates, it can be proved almost equally formally that the class does not have even one member.

the null-class is the class containing no members, not the class containing as members all unreal individuals.

It is important to observe the effect of our theory on the interpretation of definitions which proceed by means of denoting phrases. Most mathematical definitions are of this sort; for example, '$m-n$ means the number which, added to n, gives m'. Thus $m-n$ is defined as meaning the same as a certain denoting phrase; but we agreed that denoting phrases have no meaning in isolation. Thus what the definition really ought to be is: 'Any proposition containing $m-n$ is to mean the proposition which results from substituting for "$m-n$" "the number which, added to n, gives m".' The resulting proposition is interpreted according to the rules already given for interpreting propositions whose verbal expression contains a denoting phrase. In the case where m and n are such that there is one and only one number x which, added to n, gives m, there is a number x which can be substituted for $m-n$ in any proposition containing $m-n$ without altering the truth or falsehood of the proposition. But in other cases, all propositions in which '$m-n$' has a primary occurrence are false.

The usefulness of *identity* is explained by the above theory. No one outside a logic-book ever wishes to say 'x is x', and yet assertions of identity are often made in such forms as 'Scott was the author of *Waverley*' or 'thou are the man'. The meaning of such propositions cannot be stated without the notion of identity, although they are not simply statements that Scott is identical with another term, the author of *Waverley*, or that thou art identical with another term, the man. The shortest statement of 'Scott is the author of *Waverley*' seems to be 'Scott wrote *Waverley*; and it is always true of y that if y wrote *Waverley*, y is identical with Scott'. It is in this way that identity enters into 'Scott is the author of *Waverley*'; and it is owing to such uses that identity is worth affirming.

One interesting result of the above theory of denoting is this: when there is anything with which we do not have immediate acquaintance, but only definition by denoting phrases, then the propositions in which this thing is introduced by means of a denoting phrase do not really contain this thing as a constituent, but contain instead the constituents expressed by the several words of the denoting phrase. Thus in every proposition that we can apprehend (i.e. not only those whose truth or falsehood we can judge of, but in all that we can think about), all the constituents are really entities with which we have immediate acquaintance. Now such things as matter (in the sense in which matter occurs in physics) and the minds of other people are known to us only by denoting phrases, i.e. we are not *acquainted* with them, but we know them as what has such and such properties. Hence, although we can form propositional functions $C(x)$ which must hold of such and such a material particle, or of So-and-so's mind, yet we are not acquainted with the propositions which affirm these things that we know must be true, because we cannot apprehend the actual entities concerned. What we know is 'So-and-so has a mind which has such and such properties' but we do not know 'A has such and such properties', where A *is* the mind in question. In such a case, we know the properties of a thing without having acquaintance with the thing itself, and without, consequently, knowing any single proposition of which the thing itself is a constituent.

Of the many other consequences of the view I have been advocating, I will say nothing. I will only beg the reader not to make up his mind against the view – as he might be tempted to do, on account of its apparently excessive complication – until he has attempted to construct a theory of his own on the subject of denotation. This attempt, I believe, will convince him that, whatever the true theory may be, it cannot have such a simplicity as one might have expected beforehand.

12
KNOWLEDGE BY ACQUAINTANCE AND KNOWLEDGE BY DESCRIPTION

The object of the following paper is to consider what it is that we know in cases where we know propositions about "the so-and-so" without knowing who or what the so-and-so is. For example, I know that the candidate who gets most votes will be elected, though I do not know who is the candidate who will get the most votes. The problem I wish to consider is: What do we know in these cases, where the subject is merely described? I have considered this problem elsewhere[1] from a purely logical point of view; but in what follows I wish to consider the question in relation to theory of knowledge as well as in relation to logic, and in view of the above-mentioned logical discussions, I shall in this paper make the logical portion as brief as possible.

In order to make clear the antithesis between "acquaintance" and "description," I shall first of all try to explain what I mean by "acquaintance." I say that I am *acquainted* with an object when I have a direct cognitive relation to that object, i.e. when I am directly aware of the object itself. When I speak of a cognitive relation here, I do not mean the sort of relation which constitutes judgment, but the sort which constitutes presentation. In fact, I think the relation of subject and object which I call acquaintance is simply the converse of the relation of object and subject which constitutes presentation. That is, to say that S has acquaintance with O is essentially the same thing as to say that O is presented to S. But the associations and natural extensions of the word *acquaintance* are different from those of the word *presentation*. To begin with, as in most cognitive words, it is natural to say that I am acquainted with an object even at moments when it is not actually before my mind, provided it has been before my mind, and will be again whenever occasion arises. This is the same sense in which I am said to know that 2 + 2 = 4 even when I am thinking of something else. In the second place, the word *acquaintance* is designed to emphasis, more than the word *presentation*, the relational character of the fact with which we are concerned. There is, to my mind, a danger that, in speaking of presentation, we may so emphasis the object as to lose sight of the subject. The result of this is either to lead to the view that there is

---

1 See references later.

no subject, whence we arrive at materialism; or to lead to the view that what is presented is part of the subject, whence we arrive at idealism, and should arrive at solipsism but for the most desperate contortions. Now I wish to preserve the dualism of subject and object in my terminology, because this dualism seems to me a fundamental fact concerning cognition. Hence I prefer the word *acquaintance*, because it emphasises the need of a subject which is acquainted.

When we ask what are the kinds of objects with which we are acquainted, the first and most obvious example, is *sense-data*. When I see a colour or hear a noise, I have direct acquaintance with the colour or the nose. The sense-datum with which I am acquainted in these cases is generally, if not always, complex. This is particularly obvious in the case of sight. I do not mean, of course, merely that the supposed physical object is complex, but that the direct sensible object is complex and contains parts with spatial relations. Whether it is possible to be aware of a complex without being aware of its constituents is not an easy question, but on the whole it would seem that there is no reason why it should not be possible. This question arises in an acute form in connection with self-consciousness, which we must now briefly consider.

In introspection, we seem to be immediately aware of varying complexes, consisting of objects in various cognitive and conative relations to ourselves. When I see the sun, it often happens that I am aware of my seeing the sun, in addition to being aware of the sun; and when I desire food, it often happens that I am aware of my desire for food. But it is hard to discover any state of mind in which I am aware of myself alone, as opposed to a complex of which I am a constituent. The question of the nature of self-consciousness is too large, and too slightly connected with our subject, to be argued at length here. It is difficult, but probably not impossible, to account for plain facts if we assume that we do not have acquaintance with ourselves. It is plain that we are not only *acquainted* with the complex "Self-acquainted-with-A," but we also *know* the proposition "I am acquainted with A." Now here the complex has been analysed, and if "I" does not stand for something which is a direct object of acquaintance, we shall have to suppose that "I" is something known by description. If we wished to maintain the view that there is now acquaintance with Self, we might argue as follows: We are acquainted with *acquaintance*, and we know that it is a relation. Also we are acquainted with a complex in which we perceive that acquaintance is the relating relation. Hence we know that this complex must have a constituent which is that which is acquainted, i.e. must have a subject-term as well as an object-term. This subject-term we define as "I." Thus "I" means "the subject-term in awareness of which *I* am aware." But as a definition this cannot be regarded as a happy effort. It would seem necessary, therefore, either to suppose that I am acquainted with myself, and that "I," therefore, requires no definition, being merely the proper name of a certain object, or to find some other analysis of self-consciousness. Thus self-consciousness cannot be regarded as throwing light on the question whether we can know a complex without knowing its constituents. This question, however, is not important for our present purposes, and I shall therefore not discuss it further.

The awarenesses we have considered so far have all been awarenesses of particular existents, and might all in a large sense be called sense-data. For, from the point of view of theory of knowledge, introspective knowledge is exactly on a level with knowledge derived from sight or hearing. But, in addition to awareness of the above kind of objects, which may be called awareness of *particulars*, we have also (thought not quite in the same sense) what may be called awareness of *universals*. Awareness of universals is called *conceiving*, and a universal of which we are aware is called a *concept*. Not only are we aware of particular yellows, but if we have seen a sufficient number of yellows and have sufficient intelligence, we are aware of the universal *yellow*; this universal is the subject in such judgments as "yellow differs from blue" or "yellow resembles blue less than green does." And the universal yellow is the predicate in such judgments as "this is yellow," where "this" is a particular sense-datum. And universal relations, too, are objects of awarenesses; up and down, before and after, resemblance, desire, awareness itself, and so on, would seem to be all of them objects of which we can be aware.

In regard to relations, it might be urged that we are never aware of the universal relation itself, but only of complexes in which it is a constituent. For example, it may be said that we do not know directly such a relation as *before*, though we understand such a proposition as "this is before that," and may be directly aware of such a complex as "this being before that." This view, however, is difficult to reconcile with the fact that we often know propositions in which the relation is the subject, or in which the relata are not definite given objects, but "anything." For example, we know that if one thing is before another, and the other before a third, then the first is before the third; and here the things concerned are not definite things, but "anything." It is hard to see how we could know such a fact about "before" unless we were acquainted with "before," and not merely with actual particular cases of one given object being before another given object. And more directly: A judgment such as "this is before that," where this judgment is derived from awareness of a complex, constitutes an analysis, and we should not understand the analysis if we were not acquainted with the meaning of the terms employed. Thus we must suppose that we are acquainted with the meaning of "before," and not merely with instances of it.

There are thus at least two sorts of objects of which we are aware, namely, particulars and universals. Among particulars I include all existents, and all complexes of which one or more constituents are existents, such as this-before-that, this-above-that, the-yellowness-of-this. Among universals I include all objects of which no particular is a constituent. Thus the disjunctions "universal-particular" includes all objects. We might also call it the disjunction "abstract-concrete." It is not quite parallel with the opposition "concept-percept," because things remembered or imagined belong with particulars, but can hardly be called percepts. (On the other hand, universals with which we are acquainted may be identified with concepts.)

It will be seen that among the objects with which we are acquainted are not included physical objects (as opposed to sense-data), nor other people's minds. These things are known to us by what I call "knowledge by description," which we must now consider.

By a "description" I mean any phrase of the form "a so-and-so" or "the so-and-so." A phrase of the form "a so-and-so" I shall call an "ambiguous" description; a phrase of the form "the so-and-so" (in the singular) I shall call a "definite" description. Thus "a man" is an ambiguous description, and "the man with the iron mask" is a definite description. There are various problems connected with ambiguous descriptions, but I pass them by, since they do not directly concern the matter I wish to discuss. What I wish to discuss is the nature of our knowledge concerning objects in cases where we know that there is an object answering to a definite description, though we are not *acquainted* with any such object. This is a matter which is concerned exclusively with *definite* descriptions. I shall, therefore, in the sequel, speak simply of "descriptions" when I mean "definite descriptions." Thus a description will mean any phrase of the form "the so-and-so" in the singular.

I shall say that an object is "known by description" when we know that it is "*the* so-and-so," i.e. when we know that there is one object, and no more, having a certain property; and it will generally be implied that we do not have knowledge of the same object by acquaintance. We know that the man with the iron mask existed, and many propositions are known about him; but we do not know who he was. We know that the candidate who gets most votes will be elected, and in this case we are very likely also acquainted (in the only sense in which one can be acquainted with some one else) with the man who is, in fact, the candidate who will get most votes, but we do not know which of the candidates he is, i.e. we do not know any proposition of the form "A is the candidate who will get most votes" where A is one of the candidates by name. We shall say that we have "*merely* descriptive knowledge" of the so-and-so when, although we know that the so-and-so exists, and although we may possibly be acquainted with the object which is, in fact, the so-and-so, yet we do not know any proposition "*a* is the so-and-so," where *a* is something with which we are acquainted.

When we say "the so-and-so exists," we mean that there is just one object which is the so-and-so. The proposition "*a* is the so-and-so" means that *a* has the property so-and-so, and nothing else has. "Sir Joseph Larmor is the Unionist candidate" means "Sir Joseph Larmor is a Unionist candidate, and no one else is." "The Unionist candidate exists" means " some one is a Unionist candidate, and no one else is." Thus, when we are acquainted with an object which we know to be the so-and-so, we know that the so-and-so exists, but we may know that the so-and-so exists when we are not acquainted with any object which we know to be the so-and-so, and even when we are not acquainted with any object which, in fact, is the so-and-so.

Common words, even proper names, are usually really descriptions. That is to say, the thought in the mind of a person using a proper name correctly can generally only be expressed explicitly if we replace the proper name by a description. Moreover, the description required to express the thought will vary for different people, or for the same person at different times. The only thing constant (so long as the name is rightly used) is the object to which the name applies. But so long as this remains constant, the particular description involved usually makes no difference to the truth or falsehood of the proposition in which the name appears.

Let us take some illustrations. Suppose some statement made about Bismarck. Assuming that there is such a thing as direct acquaintance with oneself, Bismarck himself might have used his name directly to designate the particular person with whom he was acquainted. In this case, if he made a judgment about himself, he himself might be a constituent of the judgment. Here the proper name has the direct use which it always wishes to have, as simply standing for a certain object, and not for a description of the object. But if a person who know Bismarck made a judgment about him, the case is different. What this person was acquainted with were certain sense-data which he connected (rightly, we will suppose) with Bismarck's body. His body as a physical object, and still more his mind, were only known as the body and the mind connected with these sense-data. That is, they were known by description. It is, of course, very much a matter of chance which characteristics of a man's appearance will come into a friend's mind when he thinks of him; thus the description actually in the friend's mind is accidental. The essential point is that he knows that the various descriptions all apply to the same entity, in spite of not being acquainted with the entity in question.

When we, who did not know Bismarck, make a judgment about him, the description in our minds will probably be some more or less vague mass of historical knowledge – far more, in most cases, than is required to identify him. But, for the sake of illustration, let us assume that we think of him as "the first Chancellor of the German Empire." Here all the words are abstract except "German." The word "German" will again have different meanings for different people. To some it will recall travels in Germany, to some the look of Germany on the map, and so on. But if we are to obtain a description which we know to be applicable, we shall be compelled, at some point, to bring in a reference to a particular with which we are acquainted. Such reference is involved in any mention of past, present, and future (as opposed to definite dates), or of here and there, or of what others have told us. Thus it would seem that, in some way or other, a description known to be applicable to a particular must involve some reference to a particular with which we are acquainted, if our knowledge about the thing described is not to be merely what follows logically from the description. For example, "the most long-lived of men" is a description which must apply to some man, but we can make no judgments concerning this man which involve knowledge about him beyond what the description gives. If, however, we say, "the first Chancellor of the German Empire was an astute diplomatist," we can only be assured of the truth of our judgment in virtue of something with which we are acquainted – usually a testimony heard or read. Considered psychologically, apart from the information we convey to others, apart from the fact about the actual Bismarck, which gives importance to our judgment, the thought we really have contains the one or more particulars involved, and otherwise consists wholly of concepts. All names of places – London, England, Europe, the earth, the Solar System – similarly involve, when used, descriptions which start from some one or more particulars with which we are acquainted. I suspect that even the Universe, as considered by metaphysics, involves such a connection with particulars. In logic, on the contrary, where we are concerned not merely with what does exist, but with whatever might or could exist or be, no reference to actual particulars is involved.

It would seem that, when we make a statement about something only known by description, we often *intend* to make our statement, not in the form involving the description, but about the actual thing described. That is to say, when we say anything about Bismarck, we should like, if we could, to make the judgment which Bismarck alone can make, namely, the judgment of which he himself is a constituent. In this we are necessarily defeated, since the actual Bismarck is unknown to us. But we know that there is an object B called Bismarck, and that B was an astute diplomatist. We can thus *describe* the proposition we should like to affirm, namely, "B was an astute diplomatist," where B is the object which was Bismarck. What enables us to communicate in spite of the varying descriptions we employ is that we know there is a true proposition concerning the actual Bismarck, and that, however we may vary the description (so long as the description is correct), the proposition described is still the same. This proposition, which is described and is known to be true, is what interests us; but we are not acquainted with the proposition itself, and do not know *it*, though we know it is true.

It will be seen that there are various stages in the removal from acquaintance with particulars: there is Bismarck to people who know him, Bismarck to those who only know of him through history, the man with the iron mask, the longest-lived of men. These are progressively further removed from acquaintance with particulars, and there is a similar hierarchy in the region of universals. Many universals, like many particulars, are only known to us by description. But here, as in the case of particulars, knowledge concerning what is known by description is ultimately reducible to knowledge concerning what is known by acquaintance.

The fundamental epistemological principle in the analysis of propositions containing descriptions is this: *Every proposition which we can understand must be composed wholly of constituents with which we are acquainted.* From what has been said already, it will be plain why I advocate this principle, and how I propose to meet the case of propositions which at first sight contravene it. Let us begin with the reasons for supposing the principle true.

The chief reason for supposing the principle true is that it seems scarcely possible to believe that we can make a judgment or entertain a supposition without knowing what it is that we are judging or supposing about. If we make a judgment about (say) Julius Cæsar, it is plain that the actual person who was Julius Cæsar, is not a constituent of the judgment. But before going further, it may be well to explain what I mean when I say that this or that is a constituent of a judgment, or of a proposition which we understand. To begin with judgments: a judgment, as an occurrence, I take to be a relation of a mind to several entities, namely, the entities which compose what is judged. If, e.g. I judge that A loves B, the judgment as an event consists in the existence, at a certain moment, of a specific four-term relation, called *judging*, between me and A and love and B. That is to say, at the time when I judge, there is a certain complex whose terms are myself and A and love and B, and whose relating relation is *judging*. My reasons for this view have been set forth elsewhere,[1] and I shall not repeat them here. Assuming this view of judgment, the constituents of the judgment are simply the constituents of the complex which is the judgment. Thus, in the above case, the constituents are myself and A and love and B and judging. But myself and judging are constituents shared by all my judgments; thus the *distinctive* constituents of the particular judgment in question are A and love and B. Coming now to what is meant by "understanding a proposition," I should say that there is another relation possible between me and A and love and B, which is called my *supposing* that A loves B.[2] When we can *suppose* that A loves B, we "understand the proposition" *A loves B*. Thus we often understand a proposition in cases where we have not enough knowledge to make a judgment. Supposing, like judging, is a many term relation, of which a mind is one term. The other terms of the relation are called the constituents of the proposition supposed. Thus the principle which I enunciated may be re-stated as follows: *Whenever a relation of supposing or judging occurs, the terms to which the supposing or judging mind is related by the relation of supposing or judging must be terms with which the mind in question is acquainted.* This is merely to say that we cannot make a judgment or a supposition without knowing what it is that we are making our judgment or supposition about. It seems to me that the truth of this principle is evident as soon as the principle is understood; I shall, therefore, in what follows, assume the principle, and use it as a guide in analysing judgments that contain descriptions.

Returning now to Julius Cæsar, I assume that it will be admitted that he himself is not a constituent of any judgment which I can make. But at this point it is necessary to examine the view that judgments are composed of something called "ideas," and that it is the "idea" of Julius Cæsar that is a constituent of my judgment. I believe the plausibility of this view rests upon a failure to form a right the-

---

1 *Philosophical Essays*, "The Nature of Truth." I have been persuaded by Mr. Wittgenstein that this theory is somewhat unduly simple, but the modification which I believe it to require does not affect the above argument [1917].
2 Cf. Meinong, *Ueber Annahmen, passim*. I formerly supposed, contrary to Meinong's view, that the relationship of supposing might be merely that of presentation. In this view I now think I was mistaken, and Meinong is right. But my present view depends upon the theory that both in judgment and in assumption there is no single Objective, but the several constituents of the judgment or assumption are in a many-term relation to the mind.

ory of descriptions. We may mean by my "idea" of Julius Cæsar the things that I know about him, e.g. that he conquered Gaul, was assassinated on the Ides of March, and is a plague to schoolboys. Now I am admitting, and indeed contending, that in order to discover what is actually in my mind when I judge about Julius Cæsar, we must substitute for the proper name a description made up of some of the things I know about him. (A description which will often serve to express my thought is "the man whose name was *Julius Cæsar.*" For whatever else I may have forgotten about him, it is plain that when I mention him I have not forgotten that that was his name.) But although I think the theory that judgments consist of ideas may have been suggested in some such way, yet I think the theory itself is fundamentally mistaken. The view seems to be that there is some mental existent which may be called the "idea" of something outside the mind of the person who has the idea, and that, since judgment is a mental event, its constituents must be constituents of the mind of the person judging. But in this view ideas become a veil between us and outside things – we never really, in knowledge, attain to the things we are supposed to be knowing about, but only to the ideas of those things. The relation of mind, idea, and object, on this view, is utterly obscure, and, so far as I can see, nothing discoverable by inspection warrants the intrusion of the idea between the mind and the object. I suspect that the view is fostered by the dislike of relations, and that it is felt the mind could not know objects unless there were something "in" mind which could be called the state of knowing the object. Such a view, however, leads at once to a vicious endless regress, since the relation of idea to object will have to be explained by supposing that the idea itself has an idea of the object, and so on *ad infinitum.* I therefore see no reason to believe that, when we are acquainted with an object, there is in us something which can be called the "idea" of the object. On the contrary, I hold that acquaintance is wholly a relation, not demanding any such constituent of the mind as is supposed by advocates of "ideas." This is, of course, a large question, and one which would take us far from our subject if it were adequately discussed. I therefore content myself with the above indications, and with the corollary that, in judging, the actual objects concerning which we judge, rather than any supposed purely mental entities, are constituents of the complex which is the judgment.

When, therefore, I say that we must substitute for "Julius Cæsar" some description of Julius Cæsar, in order to discover the meaning of a judgment nominally about him, I am not saying that we must substitute an idea. Suppose our description is "the man whose name was *Julius Cæsar.*" Let our judgment be "Julius Cæsar was assassinated." Then it becomes "the man whose name was *Julius Cæsar* was assassinated." Here *Julius Cæsar* is a noise or shape with which we are acquainted, and all the other constituents of the judgment (neglecting the tense in "was") are *concepts* with which we are acquainted. Thus our judgment is wholly reduced to constituents with which we are acquainted, but Julius Cæsar himself has ceased to be a constituent of our judgment. This, however, requires a proviso, to be further explained shortly, namely that "the man whose name was *Julius*

Cæsar" must not, as a whole, be a constituent of our judgment, that is to say, this phrase must not, as a whole, have a meaning which enters into the judgment. Any right analysis of the judgment, therefore, must break up this phrase, and not treat it as a subordinate complex which is part of the judgment. The judgment "the man whose name was *Julius Cæsar* was assassinated" may be interpreted as meaning "one and only one man was called *Julius Cæsar*, and that one was assassinated." Here it is plain that there is no constituent corresponding to the phrase "the man whose name was *Julius Cæsar*." Thus there is no reason to regard this phrase as expressing a constituent of the judgment, and we have seen that this phrase must be broken up if we are to be acquainted with all the constituents of the judgment. This conclusion, which we have reached from considerations concerned with the theory of knowledge, is also forced upon us by logical considerations, which must now be briefly reviewed.

It is common to distinguish two aspects, *meaning* and *denotation*, in such phrases as "the author of *Waverley*." The meaning will be a certain complex, consisting (at least) of authorship and *Waverley* with some relation; the denotation will be Scott. Similarly "featherless bipeds" will have a complex meaning, containing as constituents the presence of two feet and the absence of feathers, while its denotation will be the class of men. Thus when we say "Scott is the author of *Waverley*" or "men are the same as featherless bipeds," we are asserting an identity of denotation, and this assertion is worth making because of the diversity of meaning.[1] I believe that the duality of meaning and denotation, though capable of a true interpretation, is misleading if taken as fundamental. The denotation, I believe, is not a constituent of the proposition, except in the case of proper names, i.e., of words which do not assign a property to an object, but merely and solely name it. And I should hold further that, in this sense, there are only two words which are strictly proper names of particulars, namely, "I" and "this."[2]

One reason for not believing the denotation to be a constituent of the proposition is that we may know the proposition even when we are not acquainted with the denotation. The proposition "the author of *Waverley* is a novelist" was known to people who did not know that "the author of *Waverley*" denoted Scott. This reason has been already sufficiently emphasised.

A second reason is that propositions concerning "the so-and-so" are possible even when "the so-and-so" has no denotation. Take, e.g. "the golden mountain does not exist" or "the round square is self-contradictory." If we are to preserve the duality of meaning and denotation, we have to say, with Meinong, that there are such objects as the golden mountain and the round square, although these objects do not have being. We even have to admit that the existent round square is exis-

---

1 This view has been recently advocated by Miss E.E.C. Jones. "A New Law of Thought and its Implications," *Mind*, January, 1911.
2 I should now exclude "I" from proper names in the strict sense, and retain only "this" [1917].

tent, but does not exist.[1] Meinong does not regard this as a contradiction, but I fail to see that it is not one. Indeed, it seems to me evident that the judgment "there is no such object as the round square" does not presuppose that there is such an object. If this is admitted, however, we are led to the conclusion that, by parity of form, no judgment concerning "the so-and-so" actually involves the so-and-so as a constituent.

Miss Jones[2] contends that there is no difficulty in admitting contradictory predicates concerning such an object as "the present King of France," on the ground that this object is in itself contradictory. Now it might, of course, be argued that this object, unlike the round square, is not self-contradictory, but merely non-existent. This, however, would not go to the root of the matter. The real objection to such an argument is that the law of contradiction ought not to be stated in the traditional form "A is not both B and not B," but in the form "no proposition is both true and false." The traditional form only applies to certain propositions, namely, to those which attribute a predicate to a subject. When the law is stated of propositions, instead of being stated concerning subjects and predicates, it is at once evident that propositions about the present King of France or the round square can form no exception, but are just as incapable of being both true and false as other propositions.

Miss Jones[3] argues that "Scott is the author of *Waverley*" asserts identity of denotation between *Scott* and *the author of Waverley*. But there is some difficulty in choosing among alternative meanings of this contention. In the first place, it should be observed that *the author of Waverley* is not a *mere* name, like *Scott*. *Scott* is merely a noise or shape conventionally used to designate a certain person; it gives us no information about that person, and has nothing that can be called meaning as opposed to denotation. (I neglect the fact, considered above, that even proper names, as a rule, really stand for descriptions.) But *the author of Waverley* is not merely conventionally a name for Scott; the element of mere convention belongs here to the separate words, *the* and *author* and *of* and *Waverley*. Given what these words stand for, *the author of Waverley* is no longer arbitrary. When it is said that Scott is the author of *Waverley*, we are not stating that these are two *names* for one man, as we should be if we said "Scott is Sir Walter." A man's name is what he is called, but however much Scott had been called the author of *Waverley*, that would not have made him be the author; it was necessary for him actually to write *Waverley*, which was a fact having nothing to do with names.

If, then, we are asserting identity of denotation, we must not mean by *denotation* the mere relation of a name to the thing named. In fact, it would be nearer to the truth to say that the *meaning* of "Scott" is the *denotation* of "the author of

---

1 Meinong, *Ueber Annahmen*, 2nd ed., Leipzig, 1910, p. 141.
2 *Mind*, July, 1910, p. 380.
3 *Mind*, July, 1910, p. 379.

Waverley." The relation of "Scott" to Scott is that "Scott" means Scott, just as the relation of "author" to the concept which is so called is that "author" means this concept. Thus if we distinguish meaning and denotation in "the author of *Waverley*," we shall have to say that "Scott" has meaning but not denotation. Also when we say "Scott is the author of *Waverley*," the *meaning* of "the author of *Waverley*" is relevant to our assertion. For if the denotation alone were relevant, any other phrase with the same denotation would give the same proposition. Thus "Scott is the author of *Marmion*" would be the same proposition as "Scott is the author of *Waverley*." But his is plainly not the case, since from the first we learn that Scott wrote *Marmion* and from the second we learn that he wrote *Waverley*, but the first tells us nothing about *Waverley* and the second nothing about *Marmion*. Hence the meaning of "the author of *Waverley*," as opposed to the denotation, is certainly relevant to "Scott is the author of *Waverley*."

We have thus agreed that "the author of *Waverley*" is not a mere name, and that its meaning is relevant in propositions in which it occurs. Thus if we are to say, as Miss Jones does, that "Scott is the author of *Waverley*" asserts an identity of denotation, we must regard the denotation of "the author of *Waverley*" as the denotation of what is *meant* by "the author of *Waverley*." Let us call the meaning of "the author of *Waverley*" M. Thus M is what "the author of *Waverley*" means. Then we are to suppose that "Scott is the author of *Waverley*" means "Scott is the denotation of M." But here we are explaining our proposition by another of the same form, and thus we have made no progress towards a real explanation. "The denotation of M," like "the author of *Waverley*," has both meaning and denotation, on the theory we are examining. If we call its meaning M′, our proposition becomes "Scott is the denotation of M′." But the leads at once to an endless regress. Thus the attempt to regard our proposition as asserting identity of denotation breaks down, and it becomes imperative to find some other analysis. When this analysis has been completed, we shall be able to reinterpret the phrase "identity of denotation," which remains obscure so long as it is taken as fundamental.

The first point to observe is that, in any proposition about "the author of *Waverley*," provided Scott is not explicitly mentioned, the denotation itself, i.e. Scott, does not occur, but only the concept of denotation, which will be represented by a variable. Suppose we say "the author of *Waverley* was the author of *Marmion*," we are certainly not saying that both were Scott – we may have forgotten that there was such a person as Scott. We are saying that there is some man who was the author of *Waverley* and the author of *Marmion*. That is to say, there is some one who wrote *Waverley* and *Marmion*, and no one else wrote them. Thus the identity is that of a variable, i.e. of an indefinite subject, "some one." This is why we can understand propositions about "the author of *Waverley*," without knowing who he was. When we say "the author of *Waverley* was a poet," we mean "one and only one man wrote *Waverley*, and he was a poet"; when we say "the author of *Waverley* was Scott" we mean "one and only one man wrote *Waverley*, and he was Scott." Here the identity is between a variable, i.e. an indeterminate subject ("he"), and Scott; "the author of *Waverley*" has been analysed away, and no longer appears as a constituent of the proposition.[1]

The reason why it is imperative to analyse away the phrase "the author of *Waverley*" may be stated as follows. It is plain that when we say "the author of *Waverley* is the author of *Marmion*," the *is* expresses identity. We have seen also that the common *denotation*, namely Scott, is not a constituent of this proposition, while the *meanings* (if any) of "the author of *Waverley*" and "the author of *Marmion*" are not identical. We have seen also that, in any sense in which the meaning of a word is a constituent of a proposition in whose verbal expression the word occurs, "Scott" means the actual man Scott, in the same sense (so far as concerns our present discussion) in which "author" means a certain universal. Thus, if "the author of *Waverley*" were a subordinate complex in the above proposition, its *meaning* would have to be what was said to be identical with the *meaning* of "the author of *Marmion*." This is plainly not the case; and the only escape is to say that "the author of *Waverley*" does not, by itself, have a meaning, though phrases of which it is part do have a meaning. That is, in a right analysis of the above proposition, "the author of *Waverley*" must disappear. This is effected when the above proposition is analysed as meaning: "Some one wrote *Waverley* and no one else did, and that some one also wrote *Marmion* and no one else did." This may be more simply expressed by saying that the propositional function "x wrote *Waverley* and *Marmion*, and no one else did" is capable of truth, i.e. some value of x makes it true, but no other value does. Thus the true subject of our judgment is a propositional function, i.e. a complex containing an undetermined constituent, and becoming a proposition as soon as this constituent is determined.

We may now define the denotation of a phrase. If we know that the proposition "a is the so-and-so" is true, i.e. that a is so-and-so and nothing else is, we call a the denotation of the phrase "the so-and-so." A very great many of the propositions we naturally make about "the so-and-so" will remain true or remain false if we substitute a for "the so-and-so," where a is the denotation of the "so-and-so." Such propositions will also remain true or remain false if we substitute for "the so-and-so" any other phrase having the same denotation. Hence, as practical men, we become interested in the denotation more than in the description, since the denotation decides as to the truth or falsehood of so many statements in which the description occurs. Moreover, as we saw earlier in considering the relations of description and acquaintance, we often wish to reach the denotation, and are only hindered by lack of acquaintance: in such cases the description is merely the means we employ to get as near as possible to the denotation. Hence it naturally comes to be supposed that the denotation is part of the proposition in which the description occurs. But we have seen, both on logical and on episte-

---

[1] The theory which I am advocating is set forth fully, with the logical grounds in its favour, in *Principia Mathematica*, Vol. I, Introduction, Chap. III: also, less fully, in *Mind*, October, 1905.

mological grounds, that this is an error. The actual object (if any) which is the denotation is not (unless it is explicitly mentioned) a constituent of propositions in which descriptions occur; and this is the reason why, in order to understand such propositions, we need acquaintance with the constituents of the description but do not need acquaintance with its denotation. The first result of analysis, when applied to propositions whose grammatical subject is "the so-and-so," is to substitute a variable as subject; i.e. we obtain a proposition of the form: "There is *something* which alone is so-and-so, and that *something* is such-and-such." The further analysis of propositions concerning "the so-and-so" is thus merged in the problem of the nature of the variable, i.e. of the meanings of *some*, *any*, and *all*. This is a difficult problem, concerning which I do not intend to say anything at present.

To sum up our whole discussion: We began by distinguishing two sorts of knowledge of objects, namely, knowledge by *acquaintance* and knowledge by *description*. Of these it is only the former that brings the object itself before the mind. We have acquaintance with sense-data, with many universals, and possibly with ourselves, but not with physical objects or other minds. We have *descriptive* knowledge of an object when we know that it is *the* object having some property or properties with which we are acquainted; that is to say, when we know that the property or properties in question belong to one object and no more, we are said to have knowledge of that one object by description, whether or not we are acquainted with the object. Our knowledge of physical objects and of other minds is only knowledge by description, the descriptions involved being usually such as involve sense-data. All propositions intelligible to us, whether or not they primarily concern things only known to us by description, are composed wholly of constituents with which we are acquainted, for a constituent with which we are not acquainted is unintelligible to us. A judgment, we found, is not composed of mental constituents called "ideas," but consists of an occurrence whose constituents are a mind[1] and certain objects, particulars or universals. (One at least must be a universal.) When a judgment is rightly analysed, the objects which are constituents of it must all be objects with which the mind which is a constituent of it is acquainted. This conclusion forces us to analyse descriptive phrases occurring in propositions, and to say that the objects denoted by such phrases are not constituents of judgments in which such phrases occur (unless these objects are explicitly mentioned). This leads us to the view (recommended also on purely logical grounds) that when we say "the author of *Marmion* was the author of *Waverley*," Scott himself is not a constituent of our judgment, and that the judgment cannot be explained by saying that it affirms identity of denotation with diversity of meaning. It also, plainly, does not assert

---

1 I use this phrase merely to denote the something psychological which enters into judgment, without intending to prejudge the question as to what this something is.

identity of meaning. Such judgments, therefore, can only be analysed by breaking up the descriptive phrases, introducing a variable, and making propositional functions the ultimate subjects. In fact, "the so-and-so is such-and-such" will mean that "x is so-and-so and nothing else is, and x is such-and-such" is capable of truth. The analysis of such judgments involves many fresh problems, but the discussion of these problems is not undertaken in the present paper.

13
LOGIC AS THE ESSENCE OF PHILOSOPHY

The topics we discussed in our first lecture,[1] and the topics we shall discuss later, all reduce themselves, in so far as they are genuinely philosophical, to problems of logic. This is not due to any accident, but to the fact that every philosophical problem, when it is subjected to the necessary analysis and purification, is found either to be not really philosophical at all, or else to be, in the sense in which we are using the word, logical. But as the word "logic" is never used in the same sense by two different philosophers, some explanation of what I mean by the word is indispensable at the outset.

Logic, in the Middle Ages, and down to the present day in teaching, meant no more than a scholastic collection of technical terms and rules of syllogistic inference. Aristotle had spoken, and it was the part of humbler men merely to repeat the lesson after him. The trivial nonsense embodied in this tradition is still set in examinations, and defended by eminent authorities as an excellent "propædeutic," i.e. a training in those habits of solemn humbug which are so great a help in later life. But it is not this that I mean to praise in saying that all philosophy is logic. Ever since the beginning of the seventeenth century, all vigorous minds that have concerned themselves with inference have abandoned the mediæval tradition, and in one way or other have widened the scope of logic.

The first extension was the introduction of the inductive method by Bacon and Galileo – by the former in a theoretical and largely mistaken form, by the latter in actual use in establishing the foundations of modern physics and astronomy. This is probably the only extension of the old logic which has become familiar to the general educated public. But induction, important as it is when regarded as a method of investigation, does not seem to remain when its work is done: in the final form of a perfected science, it would seem that everything ought to be deductive. If induction remains at all, which is a difficult question, it will remain merely as one of the principles according to which deductions are effected. Thus the ultimate result of the introduction of the inductive method seems not the creation of a new kind of non-deductive reasoning, but rather the widening of the scope of deduction by pointing out a way of deducing which is certainly not syllogistic, and does not fit into the mediæval scheme.

---

1 Editor's note: This is the second of a series of lectures, the first of which is entitled "Current Tendencies." No acquaintance with the first lecture is presupposed.

The question of the scope and validity of induction is of great difficulty, and of great importance to our knowledge. Take such a question as, "Will the sun rise tomorrow?" Our first instinctive feeling is that we have abundant reason for saying it has risen on so many previous mornings. Now, I do not myself know whether this does afford a ground or not, but I am willing to suppose that it does. The question which then arises is: "What is the principle of inference by which we pass from past sunrises to future ones?" The answer given by Mill is that the inference depends upon the law of causation. Let us suppose this to be true; then what is the reason for believing in the law of causation? There are broadly three possible answers: (1) that it is itself a known *a priori*; (2) that it is a postulate; (3) that it is an empirical generalization from past instances in which it has been found to hold. The theory that causation is known *a priori* cannot be definitely refuted, but it can be rendered very unplausible by the mere process of formulating the law exactly, and thereby showing that it is immensely more complicated and less obvious than is generally supposed. The theory that causation is a postulate, i.e. that it is something which we choose to assert although we know that it is very likely false, is also incapable of refutation; but it is plainly also incapable of justifying any use of the law in inference. We are thus brought to the theory that the law is an empirical generalization, which is the view held by Mill.

But if so, how are empirical generalizations to be justified? The evidence in their favour cannot be empirical, since we wish to argue from what has been observed to what has not been observed, which can only be done by means of some known relation of the observed and the unobserved; but the unobserved, by definition, is not known empirically, and therefore its relation to the observed, if known at all, must be known independently of empirical evidence. Let us see what Mill says on this subject.

According to Mill, the law of causation is proved by an admittedly fallible process called "induction by simple enumeration." This process, he says, "consists in ascribing the nature of general truths to all propositions which are true in every instance that we happen to know of."[1] As regards its fallibility, he asserts that "the precariousness of the method of simple enumeration is in an inverse ratio to the largeness of the generalization. The process is delusive and insufficient, exactly in proportion as the subject-matter of the observation is special and limited in extent. As the sphere widens, this unscientific method becomes less and less liable to mislead; and the most universal class of truths, the law of causation for instance, and the principles of number and of geometry, are duly and satisfactorily proved by that method alone, nor are they susceptible of any other proof."[2]

In the above statement, there are two obvious lacunæ: (1) How is the method of simple enumeration itself justified? (2) What logical principle, if any, covers the

---

1 *Logic*, Book III, chapter iii. §2.
2 Book III. chapter xxi. §3.

same ground as this method, without being liable to its failures? Let us take the second question first.

A method of proof which, when used as directed, gives sometimes truth and sometimes falsehood – as the method of simple enumeration does – is obviously not a valid method, for validity demands invariable truth. Thus, if simple enumeration is to be rendered valid, it must not be stated as Mill states it. We shall have to say, at most, that the data render the result *probable*. Causation holds, we shall say, in every instance we have been able to test; therefore it *probably* holds in untested instances. There are terrible difficulties in the notion of probability, but we may ignore them at present. We thus have what at least *may* be a logical principle, since it is without exception. If a proposition is true in every instance that we happen to know of, and if the instances are very numerous, then, we shall say, it becomes very probable, on the data, that it will be true in any further instance. This is not refuted by the fact that what we declare to be probable does not always happen, for an event may be probable on the data and yet not occur. It is, however, obviously capable of further analysis, and of more exact statement. We shall have to say something like this: that every instance of a proposition[1] being true increases the probability of its being true in a fresh instance, and that a sufficient number of favourable instances will, in the absence of instances to the contrary, make the probability of the truth of a fresh instance approach indefinitely near to certainty. Some such principle as this is required if the method of simple enumeration is to be valid.

But this brings us to our other question, namely, how is our principle known to be true? Obviously, since it is required to justify induction, it cannot be proved by induction; since it goes beyond the empirical data, it cannot be proved by them alone; since it is required to justify all inferences from empirical data to what goes beyond them, it cannot itself be even rendered in any degree probable by such data. Hence, *if* it is known, it is not known by experience, but independently of experience. I do not say that any such principle is known: I only say that it is required to justify the inferences from experience which empiricists allow, and that it cannot itself be justified empirically.[2]

A similar conclusion can be proved by similar arguments concerning any other logical principle. Thus logical knowledge is not derivable from experience alone, and the empiricist's philosophy can therefore not be accepted in its entirety, in spite of its excellence in many matters which lie outside logic.

Hegel and his followers widened the scope of logic in quite a different way – a way which I believe to be fallacious, but which requires discussion if only to show how their conception of logic differs from the conception which I wish to advocate. In their writings, logic is practically identical with metaphysics. In broad out-

---

1 Or rather a propositional function.
2 The subject of causality and induction will be discussed again in Lecture VIII.

line, the way this came about is as follows. Hegel believed that, by means of *a priori* reasoning, it could be shown that the world *must* have various important and interesting characteristics, since any world without these characteristics would be impossible and self-contradictory. Thus what he calls "logic" is an investigation of the nature of the universe, in so far as this can be inferred merely from the principle that the universe must be logically self-consistent. I do not myself believe that from this principle alone anything of importance can be inferred as regards the existing universe. But, however that may be, I should not regard Hegel's reasoning, even if it were valid, as properly belonging to logic: it would rather be an application of logic to the actual world. Logic itself would be concerned rather with such questions as what self-consistency is, which Hegel, so far as I know, does not discuss. And though he criticizes the traditional logic, and professes to replace it by an improved logic of his own, there is some sense in which the traditional logic, with all its faults, is uncritically and unconsciously assumed throughout his reasoning. It is not in the direction advocated by him, it seems to me, that the reform of logic is to be sought, but by a more fundamental, more patient, and less ambitious investigation into the presuppositions which his system shares with those of most other philosophers.

The way in which, as it seems to me, Hegel's system assumes the ordinary logic which it subsequently criticizes, is exemplified by the general conception of "categories" with which he operates throughout. This conception is, I think, essentially a product of logical confusion, but it seems in some way to stand for the conception of "qualities of Reality as a whole." Mr. Bradley has worked out a theory according to which, in all judgment, we are ascribing a predicate to Reality as a whole; and this theory is derived from Hegel. Now the traditional logic holds that every proposition ascribes a predicate to a subject, and from this it easily follows that there can be only one subject, the Absolute, for if there were two, the proposition that there were two would not ascribe a predicate to either. Thus Hegel's doctrine, that philosophical propositions must be of the form, "the Absolute is such-and-such," depends upon the traditional belief in the universality of the subject-predicate form. This belief, being traditional, scarcely self-conscious, and not supposed to be important, operates underground, and is assumed in arguments which, like the refutation of relations, appear at first sight such as to establish its truth. This is the most important respect in which Hegel uncritically assumes the traditional logic. Other less important respects – though important enough to be the source of such essentially Hegelian conceptions as the "concrete universal" and the "union of identity in difference" – will be found where he explicitly deals with formal logic.[1]

---

[1] See the translation by H.S. Macran, *Hegel's Doctrine of Formal Logic*, Oxford, 1912. Hegel's argument in this portion of his "Logic" depends throughout upon confusing the "is" of predication, as in "Socrates is mortal," with the "is" of identity, as in "Socrates is the philosopher who drank the hemlock." Owing to this confusion, he thinks that "Socrates" and "mortal" must be identical. Seeing that they are different, he does not infer, as others would, that there is a mistake

There is quite another direction in which a large technical development of logic has taken place: I mean the direction of what is called logistic or mathematical logic. This kind of logic is mathematical in two different senses: it is itself a branch of mathematics, and it is the logic which is specially applicable to other more traditional branches of mathematics. Historically, it began as *merely* a branch of mathematics: its special applicability to other branches is a more recent development. In both respects, it is the fulfilment of a hope which Leibniz cherished throughout his life, and pursued with all the ardour of his amazing intellectual energy. Much of his work on this subject has been published recently, since his discoveries have been remade by others; but none was published by him, because his results persisted in contradicting certain points in the traditional doctrine of the syllogism. We now know that on these points the traditional doctrine is wrong, but respect for Aristotle prevented Leibniz from realizing that this was possible.[1]

The modern development of mathematical logic dates from Boole's *Laws of Thought* (1854). But in him and his successors, before Peano and Frege, the only thing really achieved, apart from certain details, was the invention of a mathematical symbolism for deducing consequences from the premisses which the newer methods shared with those of Aristotle. This subject has considerable interest as an independent branch of mathematics, but it has very little to do with real logic. The first serious advance in real logic since the time of the Greeks was made independently by Peano and Frege – both mathematicians. They both arrived at their logical results by an analysis of mathematics. Traditional logic regarded the two propositions, "Socrates is mortal" and "All men are mortal," as being of the same form;[2] Peano and Frege showed that they are utterly different in form. The philosophical importance of logic may be illustrated by the fact that this confusion – which is still committed by most writers – obscured not only the whole study of the forms of judgment and inference, but also the relations of things to their qualities, of concrete existence to abstract concepts, and of the world of sense to the world of Platonic ideas. Peano and Frege, who pointed out the error, did so for technical reasons, and applied their logic mainly to technical developments; but

---

somewhere, but that they exhibit "identity in difference." Again, Socrates is particular, "mortal" is universal. Therefore, he says, since Socrates is mortal, it follows that the particular is the universal – taking the "is" to be throughout expressive of identity. But to say "the particular is the universal" is self-contradictory. Again Hegel does not suspect a mistake but proceeds to synthesize particular and universal in the individual, or concrete universal. This is an example of how, for want of care at the start, vast and imposing systems of philosophy are built upon stupid and trivial confusions, which, but for the almost incredible fact that they are unintentional, one would be tempted to characterize as puns.

1 Cf. Couturat, *La Logique de Leibniz*, pp. 361, 386.
2 It was often recognized that there was *some* difference between them, but it was not recognized that the difference is fundamental, and of very great importance.

the philosophical importance of the advance which they made is impossible to exaggerate.

Mathematical logic, even in its most modern form, is not *directly* of philosophical importance except in its beginnings. After the beginnings, it belongs rather to mathematics than to philosophy. Of its beginnings, which are the only part of it that can properly be called *philosophical* logic, I shall speak shortly. But even the later developments, though not directly philosophical, will be found of great indirect use in philosophizing. They enable us to deal easily with more abstract conceptions than merely verbal reasoning can enumerate; they suggest fruitful hypotheses which otherwise could hardly be thought of; and they enable us to see quickly what is the smallest store of materials with which a given logical or scientific edifice can be constructed. Not only Frege's theory of number, which we shall deal with in Lecture VII, but the whole theory of physical concepts which will be outlined in our next two lectures, is inspired by mathematical logic, and could never have been imagined without it.

In both these cases, and in many others, we shall appeal to a certain principle called "the principle of abstraction." This principle, which might equally well be called "the principle which dispenses with abstraction," and is one which clears away incredible accumulations of metaphysical lumber, was directly suggested by mathematical logic, and could hardly have been proved or practically used without its help. The principle will be explained in our fourth lecture, but its use may be briefly indicated in advance. When a group of objects have that kind of similarity which we are inclined to attribute to possession of a common quality, the principle in question shows that membership of the group will serve all the purposes of the supposed common quality, and that therefore, unless some common quality is actually known, the group or class of similar objects may be used to replace the common quality, which need not be assumed to exist. In this and other ways, the indirect uses of even the later parts of mathematical logic are very great; but it is now time to turn our attention to its philosophical foundations.

In every proposition and in every inference there is, besides the particular subject-matter concerned, a certain *form*, a way in which the constituents of the proposition or inference are put together. If I say, "Socrates is mortal," "Jones is angry," "The sun is hot," there is something in common in these three cases, something indicated by the word "is." What is in common is the *form* of the proposition, not an actual constituent. If I say a number of things about Socrates – that he was an Athenian, that he married Xantippe, that he drank the hemlock – there is a common constituent, namely Socrates, in all the propositions I enunciate, but they have diverse forms. If, on the other hand, I take any one of these propositions and replace its constituents, one at a time, by other constituents, the form remains constant, but no constituent remains. Take (say) the series of propositions, "Socrates drank the hemlock," "Coleridge drank the hemlock," "Coleridge drank opium," "Coleridge ate opium." The form remains unchanged throughout this series, but all the constituents have altered. Thus form is not another constituent, but is the way the constituents are put together. It is forms, in this sense, that are the proper object of philosophical logic.

It is obvious that the knowledge of logical forms is something quite different from knowledge of existing things. The form of "Socrates drank the hemlock" is not an existing thing like Socrates or the hemlock, nor does it even have that close relation to existing things that drinking has. It is something altogether more abstract and remote. We might understand all the separate words of a sentence without understanding the sentence: if a sentence is long and complicated, this is apt to happen. In such a case we have knowledge of the constituents, but not of the form. We may also have knowledge of the form without having knowledge of the constituents. If I say, "Rorarius drank the hemlock," those among you who have never heard of Rorarius, (supposing there are any) will understand the form, without having knowledge of all the constituents. In order to understand a sentence, it is necessary to have knowledge both of the constituents and of the particular instance of the form. It is in this way that a sentence conveys information, since it tells us that certain known objects are related according to a certain known form. Thus some kind of knowledge of logical forms, though with most people it is not explicit, is involved in all understanding of discourse. It is the business of philosophical logic to extract this knowledge from its concrete integuments, and to render it explicit and pure.

In all inference, form alone is essential: the particular subject-matter is irrelevant except as securing the truth of the premisses. This is one reason for the great importance of logical form. When I say "Socrates was a man, all men are mortal, therefore Socrates was mortal," the connection of premisses and conclusion does not in any way depend upon its being Socrates and man and mortality that I am mentioning. The general form of the inference may be expressed in some such words as: "If a thing has a certain property, and whatever has this property has a certain other property, then the thing in question also has that other property." Here no particular things or properties are mentioned: the proposition is absolutely general. All inferences, when stated fully, are instances of propositions having this kind of generality. If they seem to depend upon the subject-matter otherwise than as regards the truth of the premisses, that is because the premisses have not been all explicitly stated. In logic, it is a waste of time to deal with inferences concerning particular cases: we deal throughout with completely general and purely formal implications, leaving it to other sciences to discover when the hypotheses are verified and when they are not.

But the forms of propositions giving rise to inferences are not the simplest forms; they are always hypothetical, stating that if one proposition is true, then so is another. Before considering inference, therefore, logic must consider those simpler forms which inference presupposes. Here the traditional logic failed completely: it believed that there was only one form of simple proposition (i.e., of proposition not stating a relation between two or more other propositions), namely, the form which ascribes a predicate to a subject. This is the appropriate form in assigning the qualities of a given thing – we may say "this thing is round, and red, and so on." Grammar favours this form, but philosophically it is so far from universal that it is not even very common. If we say "this thing is bigger than that," we are not assigning a mere quality of "this," but a relation of "this" and "that." We might express the same fact by saying "that thing is smaller than this," where grammatically the subject is changed. Thus propositions stating that two things have a certain relation have a different form from subject-predicate propositions, and the failure to perceive this difference or to allow for it has been the source of many errors in traditional metaphysics.

The belief or unconscious conviction that all propositions are of the subject-predicate form – in other words: that every fact consists in some thing having some quality – has rendered most philosophers incapable of giving any account of the world of science and daily life. If they had been honestly anxious to give such an account, they would probably have discovered their error very quickly; but most of them were less anxious to understand the world of science and daily life, than to convict it of unreality in the interests of a super-sensible "real" world. Belief in the unreality of the world of sense arises with irresistible force in certain moods – moods which, I image, have some simple physiological basis, but are none the less powerfully persuasive. The conviction born of these moods is the source of most mysticism and of most metaphysics. When the emotional intensity of such a mood subsides, a man who is in the habit of reasoning will search for logical reasons in favour of the belief which he finds in himself. But since the belief already exists, he will be very hospitable to any reason that suggests itself. The paradoxes apparently proved by his logic are really the paradoxes of mysticism, and are the goal which he feels his logic must reach if it is to be in accordance with insight. It is in this way that logic has been pursued by those of the great philosophers who were mystics – notably Plato, Spinoza, and Hegel. But since they usually took for granted the supposed insight of the mystic emotion, their logical doctrines were presented with a certain dryness, and were believed by their disciples to be quite independent of the sudden illumination from which they sprang. Nevertheless their origin clung to them, and they remained – to borrow a useful word from Mr. Santayana – "malicious" in regard to the world of science and common sense. It is only so that we can account for the complacency with which philosophers have accepted the inconsistence of their doctrines with all the common and scientific facts which seem best established and most worthy of belief.

The logic of mysticism shows, as is natural, the defects which are inherent in anything malicious. While the mystic mood is dominant, the need of logic is not felt; as the mood fades, the impulse to logic reasserts itself, but with a desire to retain the vanishing insight, or at least to prove that it *was* insight, and that what seems to contradict it is illusion. The logic which thus arises is not quite disinterested or candid, and is inspired by a certain hatred of the daily world to which it is to be applied. Such an attitude naturally does not tend to the best results. Everyone knows that to read an author simply in order to refute him is not the way to understand him; and to read the book of Nature with a conviction that it is all illusion is just as unlikely to lead to understanding. If our logic is to find the common world intelligible, it must not be hostile, but must be inspired by a genuine acceptance such as is not usually to be found among metaphysicians.

Traditional logic, since it holds that all propositions have the subject-predicate form, is unable to admit the reality of relations: all relations, it maintains, must be reduced to properties of the apparently related terms. There are many ways of refuting this opinion; one of the easiest is derived from the consideration of what are called "asymmetrical" relations. In order to explain this, I will first explain two independent ways of classifying relations.

Some relations, when they hold between A and B, also hold between B and A. Such, for example, is the relation of "brother or sister." If A is a brother or sister of B then B is a brother or sister of A. Such again is any kind of similarity, say similarity of colour. Any kind of dissimilarity is also of this kind: if the colour of A is unlike the colour of B, then the colour of B is unlike the colour of A. Relations of this sort are called *symmetrical*. Thus a relation is symmetrical if, whenever it holds between A and B, it also holds between B and A.

All relations that are not symmetrical are called *non-symmetrical*. Thus "brother" is non-symmetrical, because, if A is a brother of B, it may happen that B is a *sister* of A.

A relation is called *asymmetrical* when, if it holds between A and B, it *never* holds between B and A. Thus husband, father, grandfather, etc., are asymmetrical relations. So are *before, after, greater, above, to the right of*, etc. All the relations that give rise to series are of this kind.

Classification into symmetrical, asymmetrical and merely non-symmetrical relations is the first of the two classifications we had to consider. The second is into transitive, intransitive, and merely none-transitive relations, which are defined as follows.

A relation is said to be *transitive*, if, whenever it holds between A and B and also between B and C, it holds between A and C. Thus *before, after, greater, above* are transitive. All relations giving rise to series are transitive, but so are many others. The transitive relations just mentioned were asymmetrical, but many transitive relations are symmetrical – for instance, equality in any respect, exact identity of colour, being equally numerous (as applied to collections), and so on.

A relation is said to be *non-transitive* whenever it is not transitive. Thus "brother" is non-transitive, because a brother of one's brother may be oneself. All kinds of dissimilarity are non-transitive.

A relation is said to be *intransitive* when, if A has the relation to B, and B to C, A never has it to C. Thus "father" is intransitive. So is such a relation as "one inch taller" or "one year later."

Let us now, in the light of this classification, return to the question whether all relations can be reduced to predications.

In the case of symmetrical relations – i.e. relations which, if they hold between

A and B, also hold between B and A – some kind of plausibility can be given to this doctrine. A symmetrical relation which is transitive, such as equality, can be regarded as expressing possession of some common property, while one which is not transitive, such as inequality, can be regarded as expressing possession of different properties. But when we come to asymmetrical relations, such as before and after, greater and less, etc., the attempt to reduce them to properties becomes obviously impossible. When, for example, two things are merely known to be unequal, without our knowing which is greater, we may say that the inequality results from their having different magnitudes, because inequality is a symmetrical relation; but to say that when one thing is *greater* than another, and not merely unequal to it, that means that they have different magnitudes, is formally incapable of explaining the facts. For if the other thing had been greater than the one, the magnitudes would also have been different, though the fact to be explained would not have been the same. Thus mere *difference* of magnitude is not *all* that is involved, since, if it were, there would be no difference between one thing being greater than another, and the other being greater than the one. We shall have to say that the one magnitude is *greater* than the *other*, and thus we shall have failed to get rid of the relation "greater." In short, both possession of the same property and possession of different properties are *symmetrical* relations, and therefore cannot account for the existence of asymmetrical relations.

Asymmetrical relations are involved in all series – in space and time, greater and less, whole and part, and many others of the most important characteristics of the actual world. All these aspects, therefore, the logic which reduces everything to subjects and predicates is compelled to condemn as error and mere appearance. To those whose logic is not malicious, such a wholesale condemnation appears impossible. And in fact there is no reason except prejudice, so far as I can discover, for denying the reality of relations. When once their reality is admitted, all *logical* grounds for supposing the world of sense to be illusory disappear. If this is to be supposed, it must be frankly and simply on the ground of mystic insight unsupported by argument. It is impossible to argue against what professes to be insight, so long as it does not argue in its own favour. As logicians, therefore, we may admit the possibility of the mystic's world, while yet, so long as we do not have his insight, we must continue to study the everyday world with which we are familiar. But when he contends that our world is impossible, then our logic is ready to repel his attack. And the first step in creating the logic which is to perform this service is the recognition of the reality of relations.

Relations which have two terms are only one kind of relations. A relation may have three terms, or four, or any number. Relations of two terms, being the simplest, have received more attention than the others, and have generally been alone considered by philosophers, both those who accepted and those who denied the reality of relations. But other relations have their importance, and are indispensable in the solution of certain problems. Jealousy, for example, is a relation between three people. Professor Royce mentions the relation "giving": when A gives B to C, that is a relation of three terms.[1] When a man says to his wife: "My dear, I wish you could induce Angelina to accept Edwin," his wish constitutes a relation between four people, himself, his wife, Angelina, and Edwin. Thus such relations are by no means recondite or rare. But in order to explain exactly how they differ from relations of two terms, we must embark upon a classification of the logical forms of facts, which is the first business of logic, and the business in which the traditional logic has been most deficient.

The existing world consists of many things with many qualities and relations. A complete description of the existing world would require not only a catalogue of things, but also a mention of all their qualities and relations. We should have to know not only this that, and the other thing, but also which was red, which yellow, which was earlier than which, which was between two others, and so on. When I speak of a "fact," I do not mean one of the simple things in the world; I mean that a certain thing has a certain quality, or that certain things have a certain relation. Thus, for example, I should not call Napoleon a fact, but I should call it a fact that he was ambitious, or that he married Josephine. Now a fact, in this sense, is never simple, but always has two or more constituents. When it simply assigns a quality to a thing, it has only two constituents, the thing and the quality. When it consists of a relation between two things, it has three constituents, the things and the relation. When it consists of a relation between three things, it has four constituents, and so on. The constituents of facts, in the sense in which we are using the word "fact," are not other facts, but are things and qualities or relations. When we say that there are relations of more than two terms, we mean that there are single facts consisting of a single relation and more than two things. I do not mean that one relation of two terms may hold between A and B, and also between A and C, as, for example, a man is the son of his father and also the son of his mother. This constitutes two distinct facts: if we choose to treat it as one fact, it is a fact which has facts for constituents. But the facts I am speaking of have no facts among their constituents, but only things and relations. For example, when A is jealous of B on account of C, there is only one fact, involving three people; there are not two instances of jealousy, but only one. It is in such cases that I speak of a relation of three terms, where the simplest possible fact in which the relation occurs is one involving three things in addition to the relation. And the same applies to relations of four terms or five or any other number. All such relations must be admitted in our inventory of the logical forms of facts: two facts involving the same number of things have the same form, and two which involve different numbers of things have different forms.

Given any fact, there is an assertion which expresses the fact. The fact itself is objective, and independent of our thought or opinion about it; but the assertion is something which involves thought, and may be either true or false. An assertion

---

[1] *Encylopædia of the Philosophical Sciences*, vol. i. p. 97.

may be positive or negative: we may assert that Charles I was executed, or that he did *not* die in his bed. The negative assertion may be said to be a *denial*. Given a form of words which must be either true or false, such as "Charles I died in his bed," we may either assert or deny this form of words: in the one case we have a positive assertion, in the other a negative one. A form of words which must be either true or false I shall call a *proposition*. Thus a proposition is the same as what may be significantly asserted or denied. A proposition which expresses what we have called a fact, i.e. which, when asserted, asserts that a certain thing has a certain quality, or that certain things have a certain relation, will be called an atomic proposition, because, as we shall see immediately, there are other propositions into which atomic propositions enter in a way analogous to that in which atoms enter into molecules. Atomic propositions, although, like facts, they may have any one of an infinite number of forms, are only one kind of propositions. All other kinds are more complicated. In order to preserve the parallelism in language as regards facts and propositions, we shall give the name "atomic facts" to the facts we have hitherto been considering. Thus atomic facts are what determine whether atomic propositions are to be asserted or denied.

Whether an atomic proposition, such as "this is red," or "this is before that," is to be asserted or denied can only be known empirically. Perhaps one atomic fact may sometimes be capable of being inferred from another, though this seems very doubtful; but in any case it cannot be inferred from premisses no one of which is an atomic fact. It follows that, if atomic facts are to be known at all, some at least must be known without inference. The atomic facts which we come to know in this way are the facts of sense-perception; at any rate, the facts of sense-perception are those which we most obviously and certainly come to know in this way. If we knew all atomic facts, and also knew that there were none except those we knew, we should, theoretically, be able to infer all truths of whatever form.[1] Thus logic would then supply us with the whole of the apparatus required. But in the first acquisition of knowledge concerning atomic facts, logic is useless. In pure logic, no atomic fact is ever mentioned: we confine ourselves wholly to forms, without asking ourselves what objects can fill the forms. Thus pure logic is independent of atomic facts; but conversely, they are, in a sense, independent of logic. Pure logic and atomic facts are the two poles, the wholly *a priori* and the wholly empirical. But between the two lies a vast intermediate region, which we must now briefly explore.

"Molecular" propositions are such as contain conjunctions – *if, or, and, unless*, etc. – and such words are the marks of a molecular proposition. Consider such an assertion as, "If it rains, I shall bring my umbrella." This assertion is just as capa-

---

1 This perhaps requires modification in order to include such facts as beliefs and wishes, since such facts apparently contain propositions as components. Such facts, though not strictly atomic, must be supposed included if the statement in the text is to be true.

ble of truth or falsehood as the assertion of an atomic proposition, but it is obvious that either the corresponding fact, or the nature of the correspondence with fact, must be quite different from what it is in the case of an atomic proposition. Whether it rains, and whether I bring my umbrella, are each severally matters of atomic fact, ascertainable by observation. But the connection of the two involved in saying that *if* the one happens, *then* the other will happen, is something radically different from either of the two separately. It does not require for its truth that it should actually rain, or that I should actually bring my umbrella; even if the weather is cloudless, it may still be true that I should have brought my umbrella if the weather had been different. Thus we have here a connection of two propositions, which does not depend upon whether they are to be asserted or denied, but only upon the second being inferable from the first. Such propositions, therefore, have a form which is different from that of any atomic proposition.

Such propositions are important to logic, because all inference depends upon them. If I have told you that if it rains I shall bring my umbrella, and if you see that there is a steady downpour, you can infer that I shall bring my umbrella. There can be no inference except where propositions are connected in some such way, so that from the truth or falsehood of the one something follows as to the truth or falsehood of the other. It seems to be the case that we can sometimes know molecular propositions, as in the above instance of the umbrella, when we do not know whether the component atomic propositions are true or false. The *practical* utility of inference rests upon this fact.

The next kind of propositions we have to consider are *general* propositions, such as "all men are mortal," "all equilateral triangles are equiangular." And with these belong propositions in which the word "some" occurs, such as "some men are philosophers" or "some philosophers are not wise." These are the denials of general propositions, namely (in the above instances), of "all men are non-philosophers" and "all philosophers are wise." We will call propositions containing the word "some" *negative* general propositions, and those containing the word "all" *positive* general propositions. These propositions, it will be seen, begin to have the appearance of the propositions in logical text-books. But their peculiarity and complexity are not known to the text-books, and the problems which they raise are only discussed in the most superficial manner.

When we were discussing atomic facts, we saw that we should be able, theoretically, to infer all other truths by logic if we knew all atomic facts and also knew that there were no other atomic facts besides those we knew. The knowledge that there are no other atomic facts is positive general knowledge; it is the knowledge that "all atomic facts are known to me," or at least "all atomic facts are in this collection" – however the collection may be given. It is easy to see that general propositions, such as "all men are mortal," cannot be known by inference from atomic facts alone. If we could know each individual man, and know that he was mortal, that would not enable us to know that all men are mortal, unless we *knew* that those were all the men there are, which is a general proposition. If we knew every other existing thing throughout the universe, and knew that each separate thing was not an immortal man, that would not give us our result unless we *knew* that we had explored the whole universe, i.e. unless we knew "all things belong to this collection of things I have examined." Thus general truths cannot be inferred from particular truths alone, but must, if they are to be known, be either self-evident or inferred from premisses of which at least one is a general truth. But all *empirical* evidence is of *particular* truths. Hence, if there is any knowledge of general truths at all, there must be *some* knowledge of general truths which is independent of empirical evidence, i.e. does not depend upon the data of sense.

The above conclusion, of which we had an instance in the case of the inductive principle, is important, since it affords a refutation of the older empiricists. They believed that all our knowledge is derived from the senses and dependent upon them. We see that, if this view is to be maintained, we must refuse to admit that we know any general propositions. It is perfectly possible logically that this should be the case, but it does not appear to be so in fact, and indeed no one would dream of maintaining such a view except a theorist at the last extremity. We must therefore admit that there is general knowledge not derived from sense, and that some of this knowledge is not obtained by inference but is primitive.

Such general knowledge is to be found in logic. Whether there is any such knowledge not derived from logic, I do not know; but in logic, at any rate, we have such knowledge. It will be remembered that we excluded from pure logic such propositions as, "Socrates is a man, all men are mortal, therefore Socrates is mortal," because Socrates and *man* and *mortal* are empirical terms, only to be understood through particular experience. The corresponding proposition in pure logic is: "If anything has a certain property, and whatever has this property has a certain other property, then the thing in question has the other property." This proposition is absolutely general: it applies to all things and all properties. And it is quite self-evident. Thus in such propositions of pure logic we have the self-evident general propositions of which we were in search.

A proposition such as "If Socrates is a man, and all men are mortal, then Socrates is mortal," is true in virtue of its *form* alone. Its truth, in this hypothetical form, does not depend upon whether Socrates actually is a man, nor upon whether in fact all men are mortal; thus it is equally true when we substitute other terms for Socrates and *man* and *mortal*. The general truth of which it is an instance is purely formal, and belongs to logic. Since this general truth does not mention any particular thing, or even any particular quality or relation, it is wholly independent of the accidental facts of the existent world, and can be known, theoretically, without any experience of particular things or their qualities and relations.

Logic, we may say, consists of two parts. The first part investigates what propositions are and what forms they may have; this part enumerates the different kinds of atomic propositions, of molecular propositions, of general propositions, and so on. The second part consists of certain supremely general propositions, which assert the truth of all propositions of certain forms. This second part merges into pure mathematics, whose propositions all turn out, on analysis, to be such general formal truths. The first part, which merely enumerates forms, is the more difficult, and philosophically the more important; and it is the recent progress in this first part, more than anything else, that has rendered a truly scientific discussion of many philosophical problems possible.

The problem of the nature of judgment or belief may be taken as an example of a problem whose solution depends upon an adequate inventory of logical forms. We have already seen how the supposed universality of the subject-predicate form made it impossible to give a right analysis of serial order, and therefore made space and time unintelligible. But in this case it was only necessary to admit relations of two terms. The case of judgment demands the admission of more complicated forms. If all judgments were true, we might suppose that a judgment consisted in apprehension of a *fact*, and that the apprehension was a relation of a mind to the fact. From poverty in the logical inventory, this view has often been held. But it leads to absolutely insoluble difficulties in the case of error. Suppose I believe that Charles I died in his bed. There is no objective fact "Charles I's death in his bed" to which I can have a relation of apprehension. Charles I and death and his bed are objective, but they are not, except in my thought, put together as my false belief supposes. It is therefore necessary, in analysing a belief, to look for some other logical form than a two-term relation. Failure to realize this necessity has, in my opinion, vitiated almost everything that has hitherto been written on the theory of knowledge, making the problem of error insoluble and the difference between belief and perception inexplicable.

Modern logic, as I hope is now evident, has the effect of enlarging our abstract imagination, and providing an infinite number of possible hypotheses to be applied in the analysis of any complex fact. In this respect it is the exact opposite of the logic practised by the classical tradition. In that logic, hypotheses which seem *prima facie* possible are professedly proved impossible, and it is decreed in advance that reality must have a certain special character. In modern logic, on the contrary, while the *prima facie* hypotheses as a rule remain admissible, others, which only logic would have suggested, are added to our stock and are very often found to be indispensable if a right analysis of the facts is to be obtained. The old logic put thought in fetters, while the new logic gives it wings. It has, in my opinion, introduced the same kind of advance into philosophy as Galileo introduced into physics, making it possible at last to see what kinds of problems may be capable of solution, and what kinds must be abandoned as beyond human powers. And where a solution appears possible, the new logic provides a method which enables us to obtain results that do not merely embody personal idiosyncrasies, but must command the assent of all who are competent to form an opinion.

14
DESCRIPTIONS

We dealt in the preceding chapter with the words *all* and *some*; in this chapter we shall consider the word *the* in the singular, and in the next chapter we shall consider the word *the* in the plural. It may be thought excessive to devote two chapters to one word, but to the philosophical mathematician it is a word of very great importance: like Browning's Grammarian with the enclitic $\delta\epsilon$, I would give the doctrine of this word if I were "dead from the waist down" and not merely in a prison.

We have already had occasion to mention "descriptive functions," *i.e.* such expressions as "the father of x" or "the sine of x." These are to be defined by first defining "descriptions."

A "description" may be of two sorts, definite and indefinite (or ambiguous). An indefinite description is a phrase of the form "a so-and-so," and a definite description is a phrase of the form "the so-and-so" (in the singular). Let us begin with the former.

"Who did you meet?" "I met a man." "That is a very indefinite description." We are therefore not departing from usage in our terminology. Our question is: What do I really assert when I assert "I met a man"? Let us assume, for the moment, that my assertion is true, and that in fact I met Jones. It is clear that what I assert is *not* "I met Jones." I may say "I met a man, but it was not Jones"; in that case, though I lie, I do not contradict myself, as I should do if when I say I met a man I really mean that I met Jones. It is clear also that the person to whom I am speaking can understand what I say, even if he is a foreigner and has never heard of Jones.

But we may go further: not only Jones, but no actual man, enters into my statement. This becomes obvious when the statement is false, since then there is no more reason why Jones should be supposed to enter into the proposition than why anyone else should. Indeed the statement would remain significant, though it could not possibly be true, even if there were no man at all. "I met a unicorn" or "I met a sea-serpent" is a perfectly significant assertion, if we know what it would be to be a unicorn or a sea-serpent, *i.e.* what is the definition of these fabulous monsters. Thus it is only what we may call the *concept* that enters into the proposition. In the case of "unicorn," for example, there is only the concept: there is not also, somewhere among the shades, something unreal which may be called "a unicorn." Therefore, since it is significant (though false) to say "I met a unicorn," it is clear that this proposition, rightly analysed, does not contain a constituent "a unicorn," though it does contain the concept "unicorn."

The question of "unreality," which confronts us at this point, is a very important one. Misled by grammar, the great majority of those logicians who have dealt with this question have dealt with it on mistaken lines. They have regarded grammatical form as a surer guide in analysis than, in fact, it is. And they have not known what differences in grammatical form are important. "I met Jones" and "I met a man" would count traditionally as propositions of the same form, but in actual fact they are of quite different forms: the first names an actual person, Jones; while the second involves a propositional function, and becomes, when made explicit: "The function 'I met x and x is human' is sometimes true." (It will be remembered that we adopted the convention of using "sometimes" as not implying more than once.) This proposition is obviously not of the form "I met x," which accounts for the existence of the proposition "I met a unicorn" in spite of the fact that there is no such thing as "a unicorn."

For want of the apparatus of propositional functions, many logicians have been driven to the conclusion that there are unreal objects. It is argued, *e.g.* by Meinong,[1] that we can speak about "the golden mountain," "the round square," and so on; we can make true propositions of which these are the subjects; hence they must have some kind of logical being, since otherwise the propositions in which they occur would be meaningless. In such theories, it seems to me, there is a failure of that feeling for reality which ought to be preserved even in the most abstract studies. Logic, I should maintain, must no more admit a unicorn than zoology can; for logic is concerned with the real world just as truly as zoology, though with its more abstract and general features. To say that unicorns have an existence in heraldry, or in literature, or in imagination, is a most pitiful and paltry evasion. What exists in heraldry is not an animal, made of flesh and blood, moving and breathing of its own initiative. What exists is a picture, or a description in words. Similarly, to maintain that Hamlet, for example, exists in his own world, namely, in the world of Shakespeare's imagination, just as truly as (say) Napoleon existed in the ordinary world, is to say something deliberately confusing, or else confused to a degree which is scarcely credible. There is only one world, the "real" world: Shakespeare's imagination is part of it, and the thoughts that he had in writing Hamlet are real. So are the thoughts that we have in reading the play. But it is of the very essence of fiction that only the thoughts, feelings, etc., in Shakespeare and his readers are real, and that there is not, in addition to them, an objective Hamlet. When you have taken account of all the feelings roused by Napoleon in writers and readers of history, you have not touched the actual man; but in the case of Hamlet

---

1 *Untersuchungen zur Gegenstandstheorie und Psychologie*, 1904.

you have come to the end of him. If no one thought about Hamlet, there would be nothing left of him; if no one had thought about Napoleon, he would have soon seen to it that some one did. The sense of reality is vital in logic, and whoever juggles with it by pretending that Hamlet has another kind of reality is doing a disservice to thought. A robust sense of reality is very necessary in framing a correct analysis of propositions about unicorns, golden mountains, round squares, and other such pseudo-objects.

In obedience to the feeling of reality, we shall insist that, in the analysis of propositions, nothing "unreal" is to be admitted. But, after all, if there *is* nothing unreal, how, it may be asked, *could* we admit anything unreal? The reply is that, in dealing with propositions, we are dealing in the first instance with symbols, and if we attribute significance to groups of symbols which have no significance, we shall fall into the error of admitting unrealities, in the only sense in which this is possible, namely, as objects described. In the proposition "I met a unicorn," the whole four words together make a significant proposition, and the word "unicorn" by itself is significant, in just the same sense as the word "man." But the *two* words "a unicorn" do not form a subordinate group having a meaning of its own. Thus if we falsely attribute meaning to these two words, we find ourselves saddled with "a unicorn," and with the problem how there can be such a thing in a world where there are no unicorns. "A unicorn" is an indefinite description which describes nothing. It is not an indefinite description which describes something unreal. Such a proposition as "x is unreal" only has meaning when "x" is a description, definite or indefinite; in that case the proposition will be true if "x" is a description which describes nothing. But whether the description "x" describes something or describes nothing, it is in any case not a constituent of the proposition in which it occurs; like "a unicorn" just now, it is not a subordinate group having a meaning of its own. All this results from the fact that, when "x" is a description, "x is unreal" or "x does not exist" is not nonsense, but is always significant and sometimes true.

We may now proceed to define generally the meaning of propositions which contain ambiguous descriptions. Suppose we wish to make some statement about "a so-and-so," where "so-and-so's" are those objects that have a certain property ϕ, *i.e.* those objects x for which the propositional function ϕx is true. (*E.g.* if we take "a man" as our instance of "a so-and-so," ϕx will be "x is human.") Let us now wish to assert the property ψ of "a so-and-so," *i.e.* we wish to assert that "a so-and-so" has that property which x has when ψx is true. (*E.g.* in the case of "I met a man," ψx will be "I met x.") Now the proposition that "a so-and-so" has the property ψ is *not* a proposition of the form "ψx." If it were, "a so-and-so" would have to be identical with x for a suitable x; and although (in a sense) this may be true in some cases, it is certainly not true in such a case as "a unicorn." It is just this fact, that the statement that a so-and-so has the property ψ is not the form of ψx, which makes it possible for "a so-and-so" to be, in a certain clearly definable sense, "unreal." The definition is as follows: –

The statement that "an object having the property ϕ has the property ψ" means:

"The joint assertion of ϕx and ψx is not always false."

So far as logic goes, this is the same proposition as might be expressed by "some ϕ's are ψ's"; but rhetorically there is a difference, because in the one case there is a suggestion of singularity, and in the other case of plurality. This, however, is not the important point. The important point is that, when rightly analysed, propositions verbally about "so-and-so" are found to contain no constituent represented by this phrase. And that is why such propositions can be significant even when there is no such thing as a so-and-so.

The definition of *existence*, as applied to ambiguous descriptions, results from what was said at the end of the preceding chapter. We say that "men exist" or "a man exists" if the propositional function "x is human" is sometimes true; and generally "a so-and-so" exists if "x is so-and-so" is sometimes true. We may put this in other language. The proposition "Socrates is a man" is no doubt *equivalent* to "Socrates is human," but it is not the very same proposition. The *is* of "Socrates is human" expresses the relation of subject and predicate; the *is* of "Socrates is a man" expresses identity. It is a disgrace to the human race that it has chosen to employ the same word "is" for these two entirely different ideas – a disgrace which a symbolic logical language of course remedies. The identity in "Socrates is a man" is identity between an object named (accepting "Socrates" as a name, subject to qualifications explained later) and an object ambiguously described. An object ambiguously described will "exist" when at least one such proposition is true, *i.e.* when there is at least one true proposition of the form "x is a so-and-so," where "x" is a name. It is characteristic of ambiguous (as opposed to definite) descriptions that there may be any number of true propositions of the above form – Socrates is a man, Plato is a man, etc. Thus "a man exists" follows from Socrates, or Plato, or anyone else. With definite descriptions, on the other hand, the corresponding form of proposition, namely, "x is the so-and-so" (where "x" is a name), can only be true for one value of x at most. This brings us to the subject of definite descriptions, which are to be defined in a way analogous to that employed for ambiguous descriptions, but rather more complicated.

We come now to the main subject of the present chapter, namely, the definition of the word *the* (in the singular). One very important point about the definition of "a so-and-so" applies equally to "the so-and-so"; the definition to be sought is a definition of propositions in which this phrase occurs, not a definition of the phrase itself in isolation. In the case of "a so-and-so," this is fairly obvious: no one could suppose that "a man" was a definite object, which could be defined by itself. Socrates is a man, Plato is a man, Aristotle is a man, but we cannot infer that "a man" means the same as "Socrates" means and also the same as "Plato" means and also the same as "Aristotle" means, since these three names have different meanings. Nevertheless, when we have enumerated all the men in the world, there is nothing left of which we can say, "This is a man, and not only so, but it is *the* 'a man,' the quintessential entity that is just an indefinite man without being anybody in particular." It is of course quite clear that whatever there is in the world is definite: if it is a man it is one definite man and not any other. Thus there cannot be such an entity as "a man" to be found in the world, as opposed to specific man. And accordingly it is natural that we do not define "a man" itself, but only the propositions in which it occurs.

In the case of "the so-and-so" this is equally true, though at first sight less obvious. We may demonstrate that this must be the case, by a consideration of the difference between a *name* and a *definite description*. Take the proposition, "Scott is the author of *Waverley*." We have here a name, "Scott," and a description, "the author of *Waverley*," which are asserted to apply to the same person. The distinction between a name and all other symbols may be explained as follows: –

A name is a simple symbol whose meaning is something that can only occur as subject, *i.e.* something of the kind that, in Chapter XIII.,[1] we defined as an "individual" or a "particular." And a "simple" symbol is one which has no parts that are symbols. Thus "Scott" is a simple symbol, because, though it has parts (namely, separate letters), these parts are not symbols. On the other hand, "the author of *Waverley*" is not a simple symbol, because the separate words that compose the phrase are parts which are symbols. If, as may be the case, whatever *seems* to be an "individual" is really capable of further analysis, we shall have to content ourselves with what may be called "relative individuals," which will be terms that, throughout the context in question, are never analysed and never occur otherwise than as subjects. And in that case we shall have correspondingly to content ourselves with "relative names." From the standpoint of our present problem, namely, the definition of descriptions, this problem, whether these are absolute names or only relative names, may be ignored, since it concerns different stages in the hierarchy of "types," whereas we have to compare such couples as "Scott" and "the author of *Waverley*," which both apply to the same object, and do not raise the problem of types. We may, therefore, for the moment, treat names as capable of being absolute; nothing that we shall have to say will depend upon this assumption, but the wording may be a little shortened by it.

We have, then, two things to compare: (1) a *name*, which is a simple symbol, directly designating an individual which is its meaning, and having this meaning in its own right, independently of the meanings of all other words; (2) a *description*, which consists of several words, whose meanings are already fixed, and from which results whatever is to be taken as the "meaning" of the description.

A proposition containing a description is not identical with what that proposition becomes when a name is substituted, even if the name names the same object

---

1 Editor's note: Chapter XIII is entitled "The Axiom of Infinity and Logical Types." The axiom of infinity, which comes up in [15], is the assumption that there exist sets with infinite cardinality; the notion of logical types is discussed in the Introduction, at section V.iii.

as the description describes. "Scott is the author of *Waverley*" is obviously a different proposition from "Scott is Scott": the first is a fact in literary history, the second a trivial truism. And if we put anyone other than Scott in place of "the author of *Waverley*," our proposition would become false, and would therefore certainly no longer be the same proposition. But, it may be said, our proposition is essentially of the same form as (say) "Scott is Sir Walter," in which two names are said to apply to the same person. The reply is that, if "Scott is Sir Walter" really means "the person named 'Scott' is the person named 'Sir Walter,'" then the names are being used as descriptions: *i.e.* the individual, instead of being named, is being described as the person having that name. This is a way in which names are frequently used in practice, and there will, as a rule, be nothing in the phraseology to show whether they are being used in this way or *as* names. When a name is used directly, merely to indicate what we are speaking about, it is no part of the *fact* asserted, or of the falsehood if our assertion happens to be false: it is merely part of the symbolism by which we express our thought. What we want to express is something which might (for example) be translated into a foreign language; it is something for which the actual words are a vehicle, but of which they are no part. On the other hand, when we make a proposition about "the person called 'Scott,'" the actual name "Scott" enters into what we are asserting, and not merely into the language used in making the assertion. Our proposition will now be a different one if we substitute "the person called 'Sir Walter.'" But so long as we are using names *as* names, whether we say "Scott" or whether we say "Sir Walter" is as irrelevant to what we are asserting as whether we speak English or French. Thus so long as names are used *as* names, "Scott is Sir Walter" is the same trivial proposition as "Scott is Scott." This completes the proof that "Scott is the author of *Waverley*" is not the same proposition as results from substituting a name for "the author of *Waverley*," no matter what name may be substituted.

When we use a variable, and speak of a propositional function, ϕx say, the process of applying general statements about x to particular cases will consist in substituting a name for the letter "x," assuming that ϕ is a function which has individuals for its arguments. Suppose, for example, that ϕx is "always true"; let it be, say the "law of identity," $x = x$. Then we may substitute for "x" any name we choose, and we shall obtain a true proposition. Assuming for the moment that "Socrates," "Plato," and "Aristotle" are names (a very rash assumption), we can infer from the law of identity that Socrates is Socrates, Plato is Plato, and Aristotle is Aristotle. But we shall commit a fallacy if we attempt to infer, without further premisses, that the author of *Waverley* is the author of *Waverley*. This results from what we have just proved, that, if we substitute a name for "the author of *Waverley*" in a proposition, the proposition we obtain is a different one. That is to say, applying the result to our present case: If "x" is a name, "$x = x$" is not the same proposition as "the author of *Waverley* is the author of *Waverley*," no matter what name "x" may be. Thus from the fact that all propositions of the form "$x = x$" are true we cannot infer, without more ado, that the author of *Waverley* is the author of *Waverley*. In fact, propositions of the form "the so-and-so is the so-and-so" are not always true: it is necessary that the so-and-so should *exist* (a term which will be explained shortly). It is false that the present King of France is the present King of France, or that the round square is the round square. When we substitute a description for a name, propositional functions which are "always true" may become false, if the description describes nothing. There is no mystery in this as soon as we realise (what was proved in the preceding paragraph) that when we substitute a description the result is not a value of the propositional function in question.

We are now in a position to define propositions in which a definite description occurs. The only thing that distinguishes "the so-and-so" from "a so-and-so" is the implication of uniqueness. We cannot speak of "*the* inhabitant of London," because inhabiting London is an attribute which is not unique. We cannot speak about "the present King of France," because there is none; but we can speak about "the present King of England." Thus propositions about "the so-and-so" always imply the corresponding propositions about "a so-and-so," with the addendum that there is not more than one so-and-so. Such a proposition as "Scott is the author of *Waverley*" could not be true if *Waverley* had never been written, or if several people had written it; and no more could any other proposition resulting from a propositional function x by the substitution of "the author of *Waverley*" for "x." We may say that "the author of *Waverley*" means "the value of x for which 'x wrote *Waverley*' is true." Thus the proposition "the author of *Waverley* was Scotch," for example, involves:

(1) "x wrote *Waverley*" is not always false;
(2) "if x and y wrote *Waverley*, x and y are identical" is always true;
(3) "if x wrote *Waverley*, x was Scotch" is always true.

These three propositions, translated into ordinary language, state:

(1) at least one person wrote *Waverley*;
(2) at most one person wrote *Waverley*;
(3) whoever wrote *Waverley* was Scotch.

All these three are implied by "the author of *Waverley* was Scotch." Conversely, the three together (but no two of them) imply that the author of *Waverley* was Scotch. Hence the three together may be taken as defining what is meant by the proposition "the author of *Waverley* was Scotch."

We may somewhat simplify these three propositions. The first and second together are equivalent to: "There is a term c such that 'x wrote *Waverley*' is true when x is c and is false when x is not c." In other words, "There is a term c such that 'x wrote *Waverley*' is always equivalent to 'x is c.'" (Two propositions are "equivalent" when both are true or both are false.) We have here, to begin with, two functions of x, "x wrote *Waverley*" and "x is c," and we form a function of c by considering the equivalence of these two functions of x for all values of x; we then proceed to assert that the resulting function of c is "sometimes true," *i.e.* that it is true for at least one value of c. (It obviously cannot be true for more than one value of c.) These two conditions together are defined as giving the meaning of "the author of *Waverley* exists."

We may now define "the term satisfying the function ϕx exists." This is the general form of which the above is a particular case. "The author of *Waverley*" is "the term satisfying the function 'x wrote *Waverley*.'" And "the so-and-so" will always involve reference to some propositional function, namely, that which defines the property that makes a thing a so-and-so. Our definition is as follows –

"The term satisfying the function ϕx exists" means:

"There is a term c such that ϕx is always equivalent to 'x is c.'"

In order to define "the author of *Waverley* was Scotch," we have still to take account of the third of our three propositions, namely, "Whoever wrote *Waverley* was Scotch." This will be satisfied by merely adding that the c in question is to be Scotch. Thus "the author of *Waverley* was Scotch" is:

"There is a term c such that (1) 'x wrote *Waverley*' is always equivalent to 'x is c,' (2) c is Scotch."

And generally: "the term satisfying ϕx satisfies ψx" is defined as meaning:

"There is a term c such that (1) ϕx is always equivalent to 'x is c,' (2) ψc is true."

This is the definition of propositions in which descriptions occur.

It is possible to have much knowledge concerning a term described, *i.e.* to know many propositions concerning "the so-and-so," without actually knowing what the so-and-so is, *i.e.* without knowing any proposition of the form "x is the so-and-so," where "x" is a name. In a detective story propositions about "the man who did the deed" are accumulated, in the hope that ultimately they will suffice to demonstrate that it was A who did the deed. We may even go so far as to say that, in all such knowledge as can be expressed in words – with the exception of "this" and "that" and a few other words of which the meaning varies on different occasions – no names, in the strict sense, occur, but what seem like names are really descriptions. We may inquire significantly whether Homer existed, which we could not do if "Homer" were a name. The proposition "the so-and-so exists" is significant, whether true or false; but if a is the so-and-so (where "a" is a name), the words "a exists" are meaningless. It is only of descriptions – definite or indefinite – that existence can be significantly asserted; for, if "a" is a name, it *must* name something: what does not name anything is not a name, and therefore, if intended to be a name, is a symbol devoid of meaning, whereas a description, like "the present King of France," does not become incapable of occurring significantly merely on the ground that it describes nothing, the reason being that it is a *complex* symbol, of which the meaning is derived from that of its constituent symbols. And so, when we ask whether Homer existed, we are using the word "Homer" as an abbreviated description: we may replace it by (say) "the author of the *Iliad* and the *Odyssey*." The same considerations apply to almost all uses of what look like proper names.

When descriptions occur in propositions, it is necessary to distinguish what may be called "primary" and "secondary" occurrences. The abstract distinction is as follows. A description has a "primary" occurrence when the proposition in which it occurs results from substituting the description for "x" in some propositional function ϕx; a description has a "secondary" occurrence when the result of substituting the description for x in ϕx gives only *part* of the proposition concerned. An instance will make this clearer. Consider "the present King of France is bald." Here "the present King of France" has a primary occurrence, and the proposition is false. Every proposition in which a description which describes nothing has a primary occurrence is false. But now consider "the present King of France is not bald." This is ambiguous. If we are first to take "x is bald," then substitute "the present King of France" for "x," and then deny the result, the occurrence of "the present King of France" has a primary occurrence and the proposition is false. Confusion of primary and secondary occurrences is a ready source of fallacies where descriptions are concerned.

Descriptions occur in mathematics chiefly in the form of *descriptive functions,* *i.e.* "the term having the relation R to y," or "the R of y" as we may say, on the analogy of "the father of y" and similar phrases. To say "the father of y is rich," for example, is to say that the following propositional function of c: "c is rich, and 'x' begat y' is always equivalent to 'x is c,'" is "sometimes true," *i.e.* is true for at least one value of c. It obviously cannot be true for more than one value.

The theory of descriptions, briefly outlined in the present chapter, is of the utmost importance both in logic and in theory of knowledge. But for purposes of mathematics, the more philosophical parts of the theory are not essential, and have therefore been omitted in the above account, which has confined itself to the barest mathematical requisites.

15
MATHEMATICS AND LOGIC

Mathematics and logic, historically speaking, have been entirely distinct studies. Mathematics has been connected with science, logic with Greek. But both have developed in modern times: logic has become more mathematical and mathematics has become more logical. The consequence is that it has now become wholly impossible to draw a line between the two; in fact, the two are one. They differ as boy and man: logic is the youth of mathematics and mathematics is the manhood of logic. This view is resented by logicians who, having spent their time in the study of classical texts, are incapable of following a piece of symbolic reasoning, and by mathematicians who have learnt a technique without troubling to inquire into its meaning or justification. Both types are now fortunately growing rarer. So much of modern mathematical work is obviously on the border-line of logic, so much of modern logic is symbolic and formal, that the very close relationship of logic and mathematics has become obvious to every instructed student. The proof of their identity is, of course, a matter of detail: starting with premises which would be universally admitted to belong to logic, and arriving by deduction at results which as obviously belong to mathematics, we find that there is no point at which a sharp line can be drawn, with logic to the left and mathematics to the right. If there are still those who do not admit the identity of logic and mathematics, we may challenge them to indicate at what point, in the successive definitions and deduction of *Principia Mathematica*, they consider that logic ends and mathematics begins. It will then be obvious that any answer must be quite arbitrary.

In the earlier chapters of this book, starting from the natural numbers, we have first defined "cardinal number" and shown how to generalise the conception of number, and have then analysed the conceptions involved in the definition, until we found ourselves dealing with the fundamentals of logic. In a synthetic, deductive treatment these fundamentals come first, and the natural numbers are only reached after a long journey. Such treatment, though formally more correct than that which we have adopted, is more difficult for the reader, because the ultimate logical concepts and propositions with which it starts are remote and unfamiliar as compared with the natural numbers. Also they represent the present frontier of knowledge, beyond which is the still unknown; and the dominion of knowledge over them is not as yet very secure.

It used to be said that mathematics is the science of "quantity." "Quantity" is a vague word, but for the sake of argument we may replace it by the word "number." The statement that mathematics is the science of number would be untrue in two different ways. On the one hand, there are recognized branches of mathematics which have nothing to do with number – all geometry that does not use co-ordinates or measurement, for example: projective and descriptive geometry, down to the point at which co-ordinates are introduced, does not have to do with number, or even with quantity in the sense of *greater* or *less*. On the other hand, through the definitions of the arithmetical operations, it has become possible to generalise much that used to be proved only in connection with numbers. The result is that what was formerly the single study of Arithmetic has now become divided into numbers of separate studies, no one of which is specially concerned with numbers. The most elementary properties of numbers are concerned with one-one relations, and similarity between classes. Addition is concerned with the construction of mutually exclusive classes respectively similar to a set of classes which are not known to be mutually exclusive. Multiplication is merged in the theory of "selections," *i.e.* of a certain kind of one-many relations. Finitude is merged in the general study of ancestral relations, which yields the whole theory of mathematical induction. The ordinal properties of the various kinds of number-series, and the elements of theory of continuity of functions and the limits of functions, can be generalised so as no longer to involve any essential reference to numbers. It is a principle, in all formal reasoning, to generalise to the utmost, since we thereby secure that a given process of deduction shall have more widely applicable results; we are, therefore, in thus generalising the reasoning of arithmetic, merely following a precept which is universally admitted in mathematics. And in thus generalising we have, in effect, created a set of new deductive systems, in which traditional arithmetic is at once dissolved and enlarged; but whether any one of these new deductive systems – for example, the theory of selections – is to be said to belong to logic or to arithmetic is entirely arbitrary, and incapable of being decided rationally.

We are thus brought face to face with the question: What is this subject, which may be called indifferently either mathematics or logic? Is there any way in which we can define it?

Certain characteristics of the subject are clear. To begin with, we do not, in this subject, deal with particular things or particular properties: we deal formally with what can be said about *any* thing or *any* property. We are prepared to say that one and one are two, but not that Socrates and Plato are two, because, in our capacity of logicians or pure mathematicians, we have never heard of Socrates and Plato. A world in which there were no such individuals would still be a world in which one and one are two. It is not open to us, as pure mathematicians or logicians, to mention anything at all, because, if we do so, we introduce something irrelevant and not formal. We may make this clear by applying it to the case of the syllogism. Traditional logic says: "All men are mortal, Socrates is a man, therefore Socrates is mortal." Now it is clear that what we *mean* to assert, to begin with, is only that the premisses imply the conclusion, not that premisses and conclusion are actually true; even the most traditional logic points out that the actual truth of the premisses is irrelevant to logic. Thus the first change to be made in the above traditional syllogism is to state it in the form: "If all men are mortal and Socrates is a man, then Socrates is mortal." We may now observe that it is intended to convey that this argument is valid in virtue of its *form*, not in virtue of the particular terms occurring in it. If we had omitted "Socrates is a man" from our premisses, we should have had a non-formal argument, only admissible because Socrates is in fact a man; in that case we could not have generalised the argument. But when, as above, the argument is *formal*, nothing depends upon the terms that occur in it. Thus we may substitute α for *men*, β for mortals, and x for Socrates, where α and β are any classes whatever, and x is any individual. We then arrive at the statement: "No matter what possible values x and α and β may have, if all α's are β's and x is an α, then x is a β"; in other words, "the propositional function 'if all α's and β and x is an α, then x is a β' is always true." Here at last we have a proposition of logic – the one which is only *suggested* by the traditional statement about Socrates and men and mortals.

It is clear that, if *formal* reasoning is what we are aiming at, we shall always arrive ultimately at statements like the above, in which no actual things or properties are mentioned; this will happen through the mere desire not to waste our time proving in a particular case what can be proved generally. It would be ridiculous to go through a long argument about Socrates, and then go through precisely the same argument again about Plato. If our argument is one (say) which holds of all men, we shall prove it concerning "x," with the hypothesis "if x is a man." With this hypothesis, the argument will retain its hypothetical validity even when x is not a man. But now we shall find that our argument would still be valid if, instead of supposing x to be a man, we were to suppose him to be a monkey or a goose or a Prime Minister. We shall therefore not waste our time taking as our premiss "x is a man" but shall take "x is an α," where α is any class of individuals, or "ϕx" where ϕ is any propositional function of some assigned type. Thus the absence of all mention of particular things or properties in logic or pure mathematics is a necessary result of the fact that this study is, as we say, "purely formal."

At this point, we find ourselves faced with a problem which is easier to state than to solve. The problem is: "What are the constituents of a logical proposition?" I do not know the answer, but I propose to explain how the problem arises.

Take (say) the proposition "Socrates was before Aristotle." Here it seems obvious that we have a relation between two terms, and that the constituents of the proposition (as well as of the corresponding fact) are simply the two terms and the relation, *i.e.* Socrates, Aristotle, and *before*. (I ignore the fact that Socrates and Aristotle are not simple; also the fact that what appear to be their names are really truncated descriptions. Neither of these facts is relevant to the present issue.) We may represent the general form of such propositions by "x R y," which may be read "x has the relation R to y." This general form may occur in logical propositions, but no particular instance of it can occur. Are we to infer that the general form itself is a constituent of such logical propositions?

Given a proposition, such as "Socrates is before Aristotle," we have certain constituents and also a certain form. But the form is not itself a new constituent; if it were, we should need a new form to embrace both it and the other constituents. We can, in fact, turn *all* the constituents of a proposition into variables, while keeping the form unchanged. This is what we do when we use such a schema as "x R y," which stands for any one of a certain class of propositions, namely, those asserting relations between two terms. We can proceed to general assertions, such as "x R y is sometimes true" – *i.e.* there are cases where dual relations hold. This assertion will belong to logic (or mathematics) in the sense in which we are using the word. But in this assertion we do not mention any particular things or particular relations; no particular things or relations can ever enter into a proposition of pure logic. We are left with pure *forms* as the only possible constituents of logical propositions.

I do not wish to assert positively that pure forms – *e.g.* the form "x R y" – do actually enter into propositions of the kind we are considering. The question of the analysis of such propositions is a difficult one, with conflicting considerations on the one side and on the other. We cannot embark upon this question now, but we may accept, as a first approximation, the view that *forms* are what enter into logical propositions as their constituents. And we may explain (though not formally define) what we mean by the "form" of a proposition as follows: –

The "form" of a proposition is that, in it, that remains unchanged when every constituent of the proposition is replaced by another.

Thus "Socrates is earlier than Aristotle" has the same form as "Napoleon is greater than Wellington," though every constituent of the two propositions is different.

We may thus lay down, as a necessary (though not sufficient) characteristic of logic or mathematical propositions, that they are to be such as can be obtained from a proposition containing no variables (*i.e.* no such words as *all, some, a, the,* etc.) by turning every constituent into a variable and asserting that the result is always true or sometimes true, or that it is always true in respect of some of the variables that the result is sometimes true in respect of the others, or any variant of these forms. And another way of stating the same thing is to say that logic (or mathematics) is concerned only with *forms*, and is concerned with them only in the way of stating that they are always or sometimes true – with all the permutations of "always" and "sometimes" that may occur.

There are in every language some words whose sole function is to indicate form. These words, broadly speaking, are commonest in languages having fewest inflections. Take "Socrates is human." Here "is" is not a constituent of the proposition, but merely indicates the subject-predicate form. Similarly in "Socrates is earlier than Aristotle," "is" and "than" merely indicate form; the proposition is the same as "Socrates precedes Aristotle," in which these words have disappeared and the form is otherwise indicated. Form, as a rule, *can* be indicated otherwise than by specific words: the order of the words can do most of what is wanted. But this principle must not be pressed. For example, it is difficult to see how we could conveniently express molecular forms of propositions (*i.e.* what we call "truth-functions") without any word at all. We saw in Chapter XIV,[1] that one word or symbol is enough for this purpose, namely, a word or symbol expressing *incompatibility*. But without even one we should find ourselves in difficulties. This, however, is not the point that is important for our present purpose. What is important for us is to observe that form may be the one concern of a general proposition, even when no word or symbol in that proposition designates the form. If we wish to speak about the form itself, we must have a word for it; but if, as in mathematics, we wish to speak about all propositions that have the form, a word for the form will usually be found not indispensable; probably in theory it is *never* indispensable.

Assuming – as I think we may – that the forms of propositions *can* be represented by the forms of the propositions in which they are expressed without any special word for forms, we should arrive at a language in which everything formal belonged to syntax and not to vocabulary. In such a language we could express *all* the propositions of mathematics even if we did not know one single word of the language. The language of mathematical logic, if it were perfected, would be such a language. We should have symbols for variables, such as "x" and "R" and "y," arranged in various ways; and the way of arrangement would indicate that something was being said to be true of all values or some values of the variables. We should not need to know any words, because they would only be needed for giving values to the variables, which is the business of the applied mathematician, not of the pure mathematician or logician. It is one of the marks of a proposition of logic that, given a suitable language, such a proposition can be asserted in such a language by a person who knows the syntax without knowing a single word of the vocabulary.

But, after all, there are words that express form, such as "is" and "than." And in every symbolism hitherto invested for mathematical logic there are symbols having constant formal meanings. We may take as an example the symbol for incompatibility which is employed in building up truth-functions. Such words or symbols may occur in logic. The question is: How are we to define them?

Such words or symbols express what are called "logical constants." Logical constants may be defined exactly as we defined forms; in fact, they are in essence the same thing. A fundamental logical constant will be that which is in common among a number of propositions, any one of which can result from any other by

---

1 Editor's note: Chapter XIV is entitled "Incompatibility and the Theory of Deduction."

substitutions of terms for one another. For example, "Napoleon is greater than Wellington" results from "Socrates is earlier than Aristotle" by the substitution of "Napoleon" for "Socrates," "Wellington" for "Aristotle," and "greater" for "earlier." Some propositions can be obtained in this way from the prototype "Socrates is earlier than Aristotle" and some cannot; those that can are those that are the form "x R y," *i.e.* express dual relations. We cannot obtain from the above prototype by term-for-term substitutions such propositions as "Socrates is human" or "the Athenians gave the hemlock to Socrates," because the first is of the subject-predicate form and the second expresses a three-term relation. If we are to have any words in our pure logical language, they must be such as express "logical constants," and "logical constants" will always either be, or be derived from, what is in common among a group of propositions derivable from each other, in the above manner, by term-for-term substitution. And this which is in common is what we call "form."

In this sense all the "constants" that occur in pure mathematics are logical constants. The number 1, for example, is derivative from propositions of the form: "There is a term c such that ϕx is true when, and only when, x is c." This is a function of ϕ, and various different propositions result from giving different values to ϕ. We may (with a little omission of intermediate steps not relevant to our present purpose) take the above function of ϕ as what is meant by "the class determined by ϕ is a unit class" or "the class determined by ϕ is a member of 1"(1 being a class of classes). In this way, propositions in which 1 occurs acquire a meaning which is derived from a certain constant logical form. And the same will be found to be the case with all mathematical constants: all are logical constants, or symbolic abbreviations whose full use in a proper context is defined by means of logical constants.

But although all logical (or mathematical) propositions can be expressed wholly in terms of logical constants together with variables, it is not the case that, conversely, all propositions that can be expressed in this way are logical. We have found so far a necessary but not a sufficient criterion of mathematical propositions. We have sufficiently defined the character of the primitive *ideas* in terms of which all the ideas of mathematics can be *defined*, but not of the primitive *propositions* from which all the propositions of mathematics can be *deduced*. This is a more difficult matter, as to which it is not yet known what the full answer is.

We may take the axiom of infinity as an example of a proposition which, though it can be enunciated in logical terms, cannot be asserted by logic to be true. All the propositions of logic have a characteristic which used to be expressed by saying that they were analytic, or that their contradictories were self-contradictory. This mode of statement, however, is not satisfactory. The law of contradiction is merely one among logical propositions; it has no special pre-eminence; and the proof that the contradictory of some proposition is self-contradictory is likely to require other principles of deduction besides the law of contradiction. Nevertheless, the characteristic of logical propositions that we are in search of is the one which was felt, and intended to be defined, by those who said that it consisted in deducibility from the law of contradiction. This characteristic, which, for the moment, we may call *tautology*, obviously does not belong to the assertion that the number of individuals in the universe is n, whatever number n may be. But for the diversity of types, it would be possible to prove logically that there are classes of n terms, where n is any finite integer; or even that there are classes of $\aleph_0$ terms. But, owing to types, such proofs, as we saw in Chapter XIII[1] are fallacious. We are left to empirical observation to determine whether there are as many as n individuals in the world. Among "possible" worlds, in the Leibnizian sense, there will be worlds having one, two, three, ... individuals. There does not even seem any logical necessity why there should be even one individual[2] – why, in fact, there should be any world at all. The ontological proof of the existence of God, if it were valid, would establish the logical necessity of at least one individual. But it is generally recognised as invalid, and in fact rests upon a mistaken view of existence – *i.e.* it fails to realise that existence can only be asserted of something described, not of something named, so that it is meaningless to argue from "this is the so-and-so" and "the so-and-so exists" to "this exists." If we reject the ontological argument, we seem driven to conclude that the existence of a world is an accident – *i.e.* it is not logically necessary. If that be so, no principle of logic can assert "existence" except under a hypothesis, *i.e.* none can be of the form "the propositional function so-and-so is sometimes true." Propositions of this form, when they occur in logic, will have to occur as hypotheses or consequences of hypotheses, not as complete asserted propositions. The complete asserted propositions of logic will all be such as affirm that some propositional function is *always* true. For example, it is always true that if p implies q and q implies r then p implies r, or that, if all α's are β's and x is an α then x is a β. Such propositions may occur in logic, and their truth is independent of the existence of the universe. We may lay it down that, if there were no universe, *all* general propositions would be true; for the contradictory of a general proposition (as we saw in Chapter XV.[3]) is a proposition asserting existence, and would therefore always be false if no universe existed.

Logical propositions are such as can be known *a priori*, without study of the actual world. We only know from a study of empirical facts that Socrates is a man, but we know the correctness of the syllogism in its abstract form (i.e. when it is stated in terms of variables) without needing any appeal to experience. This is a characteristic, not of logical propositions in themselves, but of the way in which we know them. It has, however, a bearing upon the question what their nature may be, since there are some kinds of propositions which it would be very difficult to suppose we could know without experience.

---

1 Editor's note: The Axiom of Infinity and Logical Types.
2 The primitive propositions in *Principia Mathematica* are such as to allow the inference that at least one individual exists. But I now view this as a defect in logical purity.
3 Editor's note: The subject of Chapter XV is propositional functions – i.e., functions whose output, given an appropriate argument, is a proposition. This notion is discussed in section IV.ii of the Introduction.

It is clear that the definition of "logic" or "mathematics" must be sought by trying to give a new definition of the old notion of "analytic" propositions. Although we can no longer be satisfied to define logical propositions as those that follow from the law of contradiction, we can and must still admit that they are a wholly different class of propositions from those that we come to know empirically. They all have the characteristic which, a moment ago, we agreed to call "tautology." This, combined with the fact that they can be expressed wholly in terms of variables and logical constants (a logical constant being something which remains constant in a proposition even when *all* its constituents are changed) – will give the definition of logic or pure mathematics. For the moment, I do not know how to define "tautology."[1] It would be easy to offer a definition which might seem satisfactory for a while; but I know of none that I feel to be satisfactory, in spite of feeling thoroughly familiar with the characteristic of which a definition is wanted. At this point, therefore, for the moment, we reach the frontier of knowledge on our backward journey into the logical foundations of mathematics.

We have now come to an end of our somewhat summary introduction to mathematical philosophy. It is impossible to convey adequately the ideas that are concerned in this subject so long as we abstain from the use of logical symbols. Since ordinary language has no words that naturally express exactly what we wish to express, it is necessary, so long as we adhere to ordinary language, to strain words into unusual meanings; and the reader is sure, after a time if not at first, to lapse into attaching the usual meanings to words, thus arriving at wrong notions as to what is intended to be said. Moreover, ordinary grammar and syntax is extraordinarily misleading. This is the case, *e.g.*, as regards numbers; "ten men" is grammatically the same form as "white men," so that 10 might be thought to be an adjective qualifying "men." It is the case, again, wherever propositional functions are involved, and in particular as regards existence and descriptions. Because language is misleading, as well as because it is diffuse and inexact when applied to logic (for which it was never intended), logical symbolism is absolutely necessary to any exact or thorough treatment of our subject. Those readers, therefore, who wish to acquire a mastery of the principles of mathematics, will, it is to be hoped, not shrink from the labour of mastering the symbols – a labour which is, in fact, much less than might be thought. As the above hasty survey must have made evident, there are innumerable unsolved problems in the subject, and much work needs to be done. If any student is led into a serious study of mathematical logic by this little book, it will have served the chief purpose for which it has been written.

---

1 The importance of "tautology" for a definition of mathematics was pointed out to me by my former pupil Ludwig Wittgenstein, who was working on the problem. I do not know whether he has solved it, or even whether he is alive or dead.

Sources

Gottlob Frege, "*Conceptual Notation*, Preface and Chapter 1," "On the Scientific Justification of a Conceptual Notation," and "On the Aim of the 'Conceptual Notation,'" from *Conceptual Notation and Related Articles* by Gottlob Frege, translated and edited by Terrell Ward Bynum, Oxford University Press, 1972. Copyright © 1972 by Oxford University Press. Reprinted by permission of Oxford University Press.

Gottlob Frege, "Introduction," from *The Foundations of Arithmetic* by Gottlob Frege, translated by J.L. Austin, Blackwell Publishers, 1980. (Originally published in 1950.) Reprinted by permission of Blackwell Publishers.

Gottlob Frege, "Function and Concept," "On Concept and Object," "On Sense and Reference," and "What is a Function?," from *Translations from the Philosophical Writings of Gottlob Frege*, edited by Peter Geach and Max Black, Blackwell Publishers and Rowman & Littlefield Publishers, Inc., 1980. (Originally published by Philosophical Library, 1952.) Reprinted by permission of Blackwell Publishers and Rowman & Littlefield Publishers, Inc.

Gottlob Frege, "The Thought: A Logical Inquiry," translated by A.M. Quinton and M. Quinton, with a correction by A.E. Blumberg, from *Mind*, Vol. 65, 1956. Copyright © 1956 and 1971 by Oxford University Press. Reprinted by permission of Oxford University Press.

Bertrand Russell, "Mathematics and the Metaphysicians," and "Knowledge by Acquaintance and Knowledge by Description," from *Mysticism and Logic* by Bertrand Russell, Allen & Unwin, 1917. Reprinted by permission of Routledge and the Bertrand Russell Peace Foundation.

Bertrand Russell, "On Denoting," from *Logic and Knowledge* by Bertrand Russell, edited by Robert Charles Marsh, Routledge, 1988. (Originally published by Allen & Unwin and Macmillan, 1956.) Copyright © 1956 by George Allen and Unwin Limited. Reprinted by permission of Routledge and the Bertrand Russell Peace Foundation.

Bertrand Russell, "Logic as the Essence of Philosophy," from *Our Knowledge of the External World* by Bertrand Russell, Routledge, 1993. (Originally published by The Open Court Publishing Company, 1914.) Reprinted by permission of Routledge and the Bertrand Russell Peace Foundation.

Bertrand Russell, "Descriptions," and "Mathematics and Logic," from *Introduction to Mathematical Philosophy* by Bertrand Russell, Routledge, 1993. (Originally published by Allen & Unwin and The Macmillan Company, 1919.) Copyright 1919 by George Allen and Unwin Limited. Reprinted by permission of Routledge and the Bertrand Russell Peace Foundation.